£22.50

103

Arbitration Practice in Construction Contracts

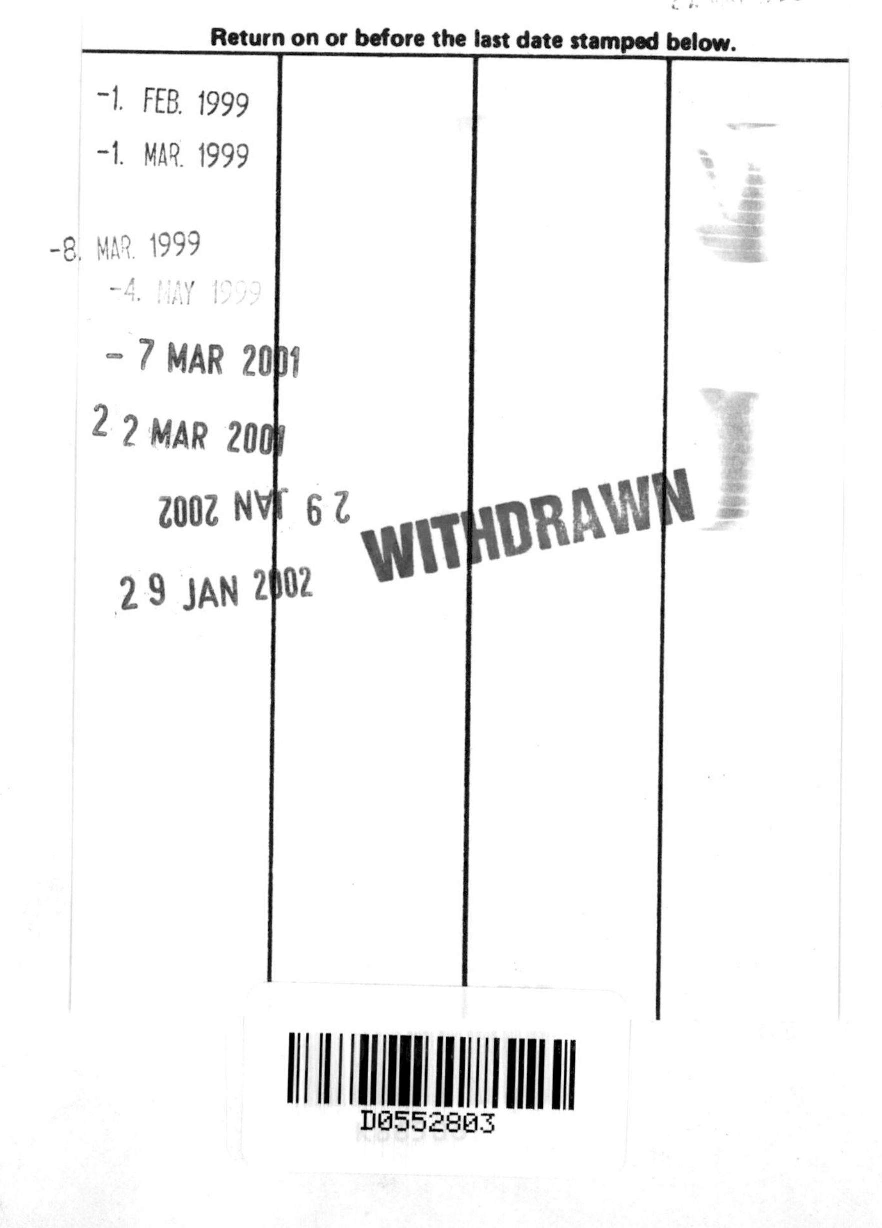

Other titles from The Builders Bookshelf Series

Builders Detail Sheets
Sam Smith
Paperback (0 419 15730 1), 200 pages

Building Regulations Explained 1992 Revision
John Stephenson
Hardback (0 419 18030 3), 550 pages

Drainage Detail Sheets
Lesley Woolley
Paperback (0 419 158804), 180 pages

Understanding JCT Standard Building Contracts
David Chappell
Paperback (0 419 18430 9), about 128 pages

For more information about these and other titles please contact:
The Promotion Department, E & FN Spon, 2–6 Boundary Row, London, SE1 8HN
Telephone 071–522–9966

Arbitration Practice in Construction Contracts

Third edition

Douglas A. Stephenson
Consulting Civil and Structural Engineer
Fellow of the Chartered Institute of Arbitrators

Foreword by The Rt Hon. Lord Goff of Chieveley

E & FN SPON
An Imprint of Chapman & Hall
London · Glasgow · New York · Tokyo · Melbourne · Madras

Published by E & FN Spon, an imprint of
Chapman & Hall, 2–6 Boundary Row, London SE1 8HN

Chapman & Hall, 2–6 Boundary Row, London SE1 8HN, UK

Blackie Academic & Professional, Wester Cleddens Road, Bishopbriggs, Glasgow G64 2NZ, UK

Chapman & Hall Inc., 29 West 35th Street, New York NY10001, USA

Chapman & Hall Japan, Thomson Publishing Japan, Hirakawacho Nemoto Building, 6F, 1–7–11 Hirakawa-cho, Chiyoda-ku, Tokyo 102, Japan

Chapman & Hall Australia, Thomas Nelson Australia, 102 Dodds Street, South Melbourne, Victoria 3205, Australia

Chapman & Hall India, R. Seshadri, 32 Second Main Road, CIT East, Madras 600 035, India

First edition 1982
Second edition 1987
Third edition 1993

Typeset in 10/12pt Plantin by Columns Design & Production Services Ltd, Reading

Printed in Great Britain by Page Bros, Norwich

ISBN 0 419 18330 2

A catalogue record for this book is available from the British Library

Library of Congress Cataloguing-in-Publication data
Stephenson, Douglas A.
Arbitration Practice in Construction Contracts / Douglas A. Stephenson. – 3rd ed.
p. cm. – (The Builders' bookshelf series)
Includes bibliographical references and index.
ISBN 0–419–18330–2 (acid-free paper)
1. Civil engineering contracts – Great Britain. 2. Construction contracts – Great Britain. 3. Arbitration and award – Great Britain.
I. Title. II. Series.
KD1641.S73 1993
343.41'078624 – dc20 93–9156
[344.10378624] CIP

∞ Printed on permanent acid-free text paper, manufactured in accordance with the proposed ANSI/NISO Z 39.48–199X and ANSI Z 39.48–1984

Contents

Appendices:

Foreword

I have been invited to contribute a Foreword to the third edition of Douglas Stephenson's book *Arbitration Practice in Construction Contracts*, and I am delighted to have been given this opportunity.

Despite the title of the book, it is evident from its contents (and indeed is stated by the author in his Preface) that the objective of the book is to provide an introduction to the practice of arbitration in general. There is, of course, a certain emphasis on building and construction, as for example in the account of the standard forms in chapter 2, and in the concluding chapter which is devoted to the contractor as claimant. But the greater part of the book is designed to lead readers gently through the various stages of an arbitration, at each stage placing them in their legal context so that they can understand why the applicable law and procedure takes the form it does, and how it actually works in practice.

Obviously, the book is modest in size. But it is essentially a handbook for those, especially civil engineering contractors, who find themselves faced with the impact of arbitration law and practice upon their practical work. For this purpose the scale of the book is one of its great virtues. Readers are not drowned in detail, as they tend to be when consulting standard legal textbooks. On the contrary, simplicity, economy and clarity are the hallmarks of this book; and it is the combination of these three qualities which will ensure that readers will rapidly find what they want; that they will never lose their way; and that they will not be distracted by unnecessary detail. After all, if a reader wants to pursue a point further, he or she can then turn to one of the standard textbooks for help.

The usefulness of the book is enhanced by the inclusion of some very helpful examples of specimen documents, without which it must be very difficult for those unaccustomed to legal procedures to understand the shape which such documents usually take, or the purposes which they are designed to perform. In addition, the relevant statutory provisions are set out to be consulted where necessary.

All in all, this is a book which is designed to provide the reader with as much practical information as possible about arbitration law and procedure, in an accessible form, and within a modest compass. The author has surely succeeded admirably in fulfilling this thoroughly useful purpose.

The Rt Hon. Lord Goff of Chieveley
March 1993

Preface to the third edition

During the 11 years that have elapsed since publication of the first edition I have received much helpful (and some critical) comment from readers. I hasten to add that the criticism has been entirely constructive, and is most welcome. It has enabled me to correct a few errors in, and omissions from, the text, and to expand those parts of the book where further guidance has been requested. In particular, the Specimen Documents have been a subject of praise and of requests for expansion, and in this edition I have revised those documents extensively in order to incorporate recent procedural developments and to include additional forms.

The past decade has also seen many changes in both statute and common law of arbitration. There has been a welcome clarification, in section 19A of the 1950 Act, of the arbitrator's power to award interest, albeit that it is expressly *simple* interest. However, the contractual right to *compound* interest under the fifth edition of the ICE Conditions has also been clarified, in case law, and this has been closely followed by a corresponding amendment incorporated in the sixth edition of that document. The widely criticized decision on the House of Lords in *Bremer Vulkan* case, that an arbitrator is not empowered to dismiss a claim for want of prosecution, has been the subject of a statutory reversal under the Courts and Legal Services Act 1990. The House of Lords' decision in *The Nema* has been supplemented and clarified by their Lordships' decision in *The Antaios*, resulting in clearly defined rules governing applications for leave to appeal under section 1 of the 1979 Act.

Alternative Dispute Resolution in its various forms has gained in popularity during the past few years: it remains to be seen whether the benefits resulting from formalizing the traditional 'horse trading' between employers and contractors are real or illusory. Certainly any initiative which may result in settlements being achieved sooner rather than later is to be welcomed.

But perhaps the most important of recent developments has been the growing confidence among arbitrators that the courts will not overturn their awards without good reason: a confidence which has encouraged more robust treatment of parties who seek to cause delay or to frustrate proceedings by legalistic contentions.

These wholly welcome developments have helped to enhance the reputation of arbitration as a just, expeditious and economical means of dispute resolution. They have also emphasized the need for a further edition of this book which has I believe achieved a modest degree of success in promoting efficient arbitration procedures, and will I hope continue to do so.

In 1986 the Rt Hon. Lord Goff of Chieveley, having completed his meteoric rise from the High Court to the House of Lords (during which I suspect he overtook many cases that were making the same ascent) became President of the Chartered Institute of Arbitrators. He held that office for a record term of five years, and during the first of those years I had the privilege of serving under him as Chairman of the Institute.

It would be difficult indeed to find anyone better qualified to write a Foreword for this book, and I am delighted and honoured that Lord Goff agreed to do so. It is an added bonus that he has written it in the concise and forthright terms that are so evident in his judgments, but perhaps with more than a touch of generosity.

Douglas A. Stephenson
March 1993

Preface to the first edition

Although written primarily as a guide to the civil engineering contractor who finds himself involved in a dispute arising from a construction contract this book necessarily covers a wider field than that seen by one of the parties. For if arbitration is to service its purpose as an inexpensive and expeditious means of resolving such disputes it must be properly understood and used; and understanding cannot be gained through a blinkered study from a single viewpoint. Accordingly the objective of this book is to provide an introduction to the practice of arbitration in general. I have, however, chosen examples relating to (imaginary) construction disputes.

A by-product of this more general approach to the subject is that the book will, I hope, be of value not only to contractors in the narrow sense of the word but also to employers, engineers and arbitrators in the field of construction, and to those who are contractors in the widest sense: namely that they are parties to a commercial contract.

Although arbitration has for many decades been the chosen method by which disputes arising from construction contracts are resolved, the manner of its use, and in particular the way in which arbitrators have been chosen, has until recently left much to be desired. It is only since the early 1970s that the need for arbitrators to know about arbitration has been recognized by the Institution of Civil Engineers; prior to this the principal, if not the sole, criterion was eminence as a civil engineer. Too often the arbitrator appointed was in the position of that person who is reputed, when asked if he could play the piano, to have replied: 'I don't know – I've never tried.'

Inevitably one consequence of this lack of understanding has been to turn for advice to lawyers, many of whom see arbitration as a poor substitute for legal action through the courts, with all of the disadvantages associated with such actions but none of the advantages. Arbitration has in this way gained a reputation for being anything but inexpensive and expeditious; and the procedure has been blamed for shortcomings for which the real blame lies with its users.

Much has been done by the Chartered Institute of Arbitrators to foster understanding and efficient use of the procedure, by means of training courses and by setting standards of proficiency as a prerequisite to appointment to panels from which arbitrators are selected when the Chartered Institute is the appointing authority.

My experience during the past few years as a tutor and more recently as a course director of certain of these training courses has proved of immense value in writing this book. In particular it has drawn attention to the difficulty encountered by many students in drafting the many letters and documents used in the proceedings, notwithstanding that these usually follow a standard format requiring little adaptation to suit a particular reference. Accordingly I have included, in Appendix A, a series of specimen letters and documents which, while not applicable to every situation, will in most cases provide at least a format and a basis for adaptation to a particular need.

The reasons underlying the choice of arbitration as the means whereby construction and other disputes are resolved are not hard to find. It is flexible, enabling it to be adapted to suit the needs of a particular dispute, large or small, simple or complex. It provides for determination by a person who understands the technicalities of the subject matter, and for the privacy and convenience of the parties. And properly used it should be both cheaper and quicker than litigation.

Where the parties, under the direction of a sensible arbitrator, behave sensibly, all of these objectives should be achieved. Where one party finds itself confronted by an obstructive or dilatory opponent it should bring such behaviour to the notice of the arbitrator with a request that it be taken into account in his award of costs. And where a party finds the arbitrator to be inexperienced it can lead him gently to adopt sensible procedures.

If this book helps in promoting the use of such procedures it will have served its purpose.

Being of an introductory nature this small volume does not attempt to deal with the complexities of the law or with procedural difficulties that sometimes arise during the course of a reference. For such matters 'Russell' and other major works are available and are listed in the bibliography. I have however attempted to explain in simple terms the significance of some of the more important decisions of the courts affecting the conduct of arbitrations.

In writing the Foreword Lord Justice Donaldson, President of the Chartered Institute of Arbitrators, has bestowed upon this book an honour that I can only hope it deserves. He has also read through the typescript and given guidance on two matters of legal complexity, namely the law relating to the award of interest, and the powers available to an arbitrator, through the High Court, in dealing with reluctant or obstructive parties. For both of these actions, and for his encouragement, I am deeply grateful.

I am also indebted to my good friend and co-tutor Ian Menzies for his preliminary reading of the typescript and for making many helpful and constructive suggestions.

Douglas A. Stephenson
March 1982

Acknowledgements

The author and publishers wish to acknowledge permission from the following to reproduce copyright material: Her Majesty's Stationery Office (Appendices B, C, D, E, F); The Institution of Civil Engineers (Form ArbICE (revised): The Society of Construction Arbitrators (Form of Agreement appended to SD/9).

1 Introduction

SYNOPSIS

The basic principle of arbitration, namely that the parties to a contract from which a dispute arises elect to appoint a tribunal of their own choice to determine that dispute, is especially relevant where, as in construction, technicalities are involved. But arbitration depends for its efficacy upon the law: for without a framework of law within which recognition is given, the award of an arbitrator could prove worthless if the losing party chose to ignore it. Enforcement of arbitration awards is available to the parties through the courts of England and of most other civilized countries: though the law of England will in general require that the arbitration proceedings have been conducted, and the award made, in accordance with that law.

DEFINITION

'An arbitrator is a private extraordinary judge between party and party, chosen by their mutual consent to determine controversies between them. And arbitrators are so called because they have an arbitrary power: for if they observe the submission and keep within due bounds, their sentences are definite from which there lies no appeal.'

The words of Lord Chief Justice Sir Robert Raymond, expressed some 250 years ago, still provide a valid definition: for if the phrase 'due bounds' is taken to mean 'the law', there is indeed no appeal from an arbitrator's award. The limited grounds upon which an appeal may lie only cover points of law, including failure by the arbitrator to conduct himself or the proceedings in accordance with the law (see Chapter 10).

Arbitration is a voluntary procedure, available as an alternative to litigation, but not enforceable as the means of settling disputes except where the parties have entered into an arbitration agreement. In such cases the right of either party to have disputes resolved by arbitration will, except where there are good reasons to the contrary, be upheld by the court (see p. 9).

LEGAL FRAMEWORK

This book is concerned with arbitration under English law, which applies in England and in Wales, but not in Scotland or in Northern Ireland. English law applies automatically as the procedural law of arbitrations arising from contracts under the ICE Conditions, the FCEC Form of Subcontract, the JCT Forms of Contract or any other standard form where the contract is made and the work is to be performed in England or Wales, and where no other system of law is specified in the contract.

English law will also apply to arbitrations arising from contracts of a non-domestic type (that is, where one or both parties are based overseas, or where the work is to be performed overseas), where the parties elect that it shall apply. Their choice in this matter is not necessarily affected by their choice of the proper law of the contract (the law by which the contract is to be construed), although in general there will be a presumption to that effect where no other system of law is specified.

It is desirable that where English law governs the arbitration procedure the place of the arbitration should be within the jurisdiction of English courts so that the many 'supervisory' functions provided by those courts and described in this book are readily available.

There are indeed good reasons why arbitrations should be conducted under English law even in the case of non-domestic contracts, and even when the proper law of the contract is not English law. Firstly, English commercial law, and especially arbitration law, is more highly developed and sophisticated than any other legal system. Secondly, it forms the basis of many other legal systems throughout the world, and is therefore more readily accepted and understood than other systems. Thirdly, the general use of the English language in many overseas contracts reduces the problems of interpretation where English arbitration law is adopted.

ENGLISH LAW OF ARBITRATION

Arbitration in England is known to have been recognized in common law since the beginning of the seventeenth century: indeed the system is claimed to be as old as legal history. The first statute was the Arbitration Act 1697, since which date there have been a number of re-enactments. Fortunately for today's practitioners, however, much of the law of arbitration was summarized and codified in the Arbitration Act 1950 (referred to herein as the 1950 Act) and this Act, except for one major change, has remained substantially unaltered, being now referred to as the 'principal Act'.

The 1950 Act (see Appendix B) has the merits of simplicity and clarity. It also performs the valuable function of filling gaps that might otherwise exist in arbitration agreements, by defining the constitution of the tribunal, authorizing the arbitrator to make orders, to make an award at any time, to make an interim award, to order specific performance and to award costs. But perhaps the most important function of the 1950 Act is that contained in section 26 wherein an arbitration award may effectively be converted into a judgment of the High Court.

One of the main features of arbitration under English law is the emphasis laid upon compliance with the law. The danger has long been recognized that arbitrators, who are generally not qualified legally and who conduct proceedings in private, may stray from the law of the land, developing their own fund of 'case law' based upon their own concept of equity. In so doing they might well diverge from the law, not only as a body of arbitrators, but also as individuals; 'legal' precedents set by each individual might differ both from those of common law and from each other. In order to control this danger a system was set up, originally under the Common Law Procedure Act 1854 and more recently in the 1950 Act, whereby a point of law arising either during the course of a reference or in an award might, on the application of either party or of the arbitrator, be made the subject of a 'Special Case' for determination by the High Court.

Although this procedure, which was defined in section 21 (later repealed) of the 1950 Act, provided a means by which case law developed to a high degree of sophistication in commercial contracts, it also provided a means whereby an unmeritorious party might delay the day of judgment. For by applying to the arbitrator for an award in the form of a special case on some spurious point of law, and then pursuing that point through the hierarchy of the courts, the losing party in an arbitration could delay, sometimes for many years, the date upon which payment became due, thereby gaining a financial advantage because of the unrealistically low rate of interest formerly allowed in law upon arbitration awards.

Pressure from a number of commercial and legal bodies, and in particular from the Chartered Institute of Arbitrators, has resulted in the special case procedure being repealed, and replaced by a limited right of appeal on points of law, under section 1 of the Arbitration Act 1979 (referred to herein as the 1979 Act; see Appendix D), which came into force on 1 August 1979 and applies to all arbitrations commenced after that date, and to other arbitration where it is adopted by agreement of the parties.

The 1979 Act makes provision, under section 2, for a limited right to apply to the High Court for determination of a point of law arising in the course of a reference (see Chapter 10). It also fills certain gaps in the principal Act, such as that which arises when an appointing authority named in an arbitration agreement neglects or refuses to make the appointment, and it strengthens the power of an arbitrator to deal with a recalcitrant party.

In order to complete the list of statutes currently applicable to arbitration mention is also necessary of the Arbitration Act 1975 (Appendix C), the principal purpose of which is 'to give effect to the New York

Convention on the Recognition and Enforcement of Foreign Arbitral Awards'. This Act is not, however, relevant to domestic arbitrations.

ADVANTAGES OF ARBITRATION

Many of the advantages most frequently claimed of arbitration as an alternative to litigation are especially relevant to those arising from construction contracts.

Freedom in choice of arbitrator

The parties to an arbitration agreement are free to choose a suitable person to be arbitrator. Frequently disputes arising from construction contracts involve such questions as whether or not certain ground conditions could reasonably have been foreseen by an experienced contractor; whether or not it was reasonable for the engineer or architect to issue drawings when he did, or to give the instructions he gave; or questions may involve the technicalities of the various standard methods of measurement. A proper understanding of these and many other points that may arise can only be gained by long experience in the construction industry – and preferably experience both in contracting and as the engineer under the contract. Hence it is desirable that the arbitrator should in such cases be an experienced engineer (or, where appropriate, architect or quantity surveyor): and this objective is usually best achieved by naming as the appointed authority the president of the appropriate professional body.

While it is recognized that technical expertise is available in litigation through the appointment of experts, there is a very real danger that a non-technical judge may be influenced more by the eloquence and powers of explanation and persuasion of the expert than by the technical merits of his evidence.

Flexibility

Disputes arising from construction contracts may in some cases involve a few thousand – perhaps even a few hundred – pounds, or they may involve tens or hundreds of millions. They may involve questions of principle that will affect future contracts, or may (more commonly) be of a 'one-off' nature affecting only the contract from which the dispute arose. They may involve technical or legal issues or both: and the credibility of the witnesses to be called may or may not be in doubt. All of these factors affect the choice of procedure and the most appropriate form and level of representation – if any – of the parties. In arbitration the parties are free to determine these matters by agreement; and while neither party can dictate to the other where it is thought, for example, that costs are being incurred unnecessarily, the party may bring its contentions to the notice of the arbitrator and request that he take them into consideration in his award of costs.

Economy

Critics of arbitration often argue that costs in total are likely to exceed those incurred in litigation, because in the latter the judge is paid out of the public purse, while the arbitrator is not. While true, this is not usually a major factor in the total costs of the proceedings: the arbitrator's charges are often much less than those of the parties' solicitors and counsel. Furthermore, where technical matters are involved it is likely that experts will be needed to present such matters to a judge, but not to a technically qualified arbitrator. Again, proceedings in court are likely to be more protracted, and hence more costly, than in arbitration.

Economy is not achieved automatically by use of arbitration rather than litigation, but it may be achieved where the parties act sensibly in choosing the form of the proceedings and of their representation. Where one party acts sensibly but the other does not, it is within the power of the arbitrator to award costs accordingly.

Expedition

It is especially important in construction disputes that unnecessary delay in their resolution should be avoided. Such disputes often involve voluminous documentary and oral evidence of details of progress, instructions, delaying factors and other matters. With the passage of time records may get lost or dispersed; memories fade; staff move on or die; defective work may become obscured or affected by alterations. Furthermore, in some cases progress of the remaining work may be delayed pending resolution of a dispute, for example where a contractor encounters conditions which he had not foreseen and which he claims entitle him to reimbursement of extra costs under Clause 12 of the ICE Conditions. Generally arbitration, if properly used, provides the means of resolving a dispute with the minimum of delay.

Privacy

Arbitration proceedings, unlike those in the courts, are not open to the press or to the public; only those persons involved in the proceedings are entitled to attend. It is of course open to the parties, by agreement, to allow others to be present, and they often do so where for example the arbitrator wishes a pupil to gain experience. Such attendance is normally on the condition that the confidentiality of the proceedings will be respected.

Usually the parties to construction disputes have no wish to publicize either the matters disclosed at the hearing or any other details of a reference; and frequently the damage to a previously harmonious relationship between two parties resulting from a dispute is more rapidly healed where there has been no publicity.

Finality

One respect in which arbitration under English law differs from that under most other legal systems is in the emphasis given to the need for awards to comply with the law. This point has been the subject of much criticism and debate in recent years, it being argued against the English system that finality is more important than legality and that, having chosen their arbitrator, the parties should be content to accept his decision, whether strictly in accordance with legal precedent or not.

Until the 1979 Act became law parties were able on the flimsiest of pretexts to ensure that an award, when given, would not be final, but was capable of being challenged through the courts. Now, however, the right to appeal is strictly limited (see Chapter 10) and may in some circumstances be eliminated entirely.

DISADVANTAGES OF ARBITRATION

Cost of arbitrator and of court facilities

In contrast to litigation, wherein both the judge and the court facilities are provided at public expense, the parties to an arbitration, or one of them, will ultimately have to bear the costs of the arbitrator and of the courtroom facilities. However, in most cases such costs are small in comparison with other costs incurred in litigation (see p. 5).

Legal aid is not available

Where arbitration is used as a means of resolving minor disputes, and in particular those in consumer industries, the non-availability of legal aid may be an important consideration. However, in many such arbitration schemes administered by the Chartered Institute of Arbitrators the aim is to obviate any need for legal representation, and to keep costs to a nominal amount.

Joinder difficulties

Where more than two parties are involved in a dispute – for example employer, main contractor and subcontractor – there is no statutory power whereby all parties may, as in litigation, be joined in a consolidated action. Certain standard forms of contract, such as the FCEC Form of Subcontract and the JCT Form of Building Contract, do however provide for consolidation in limited circumstances.

Incompetent arbitrators

While judges are appointed only after they have gained extensive experience, usually at the bar, arbitrators having inadequate qualification may be, and sometimes are, appointed either by an appointing authority or by the parties, in ignorance of the requirements of the appointment.

Many appointing authorities now maintain panels of qualified arbitrators; often requiring candidates for listing on those panels to qualify at the Chartered Institute of Arbitrators before taking the professional body's own examination. Parties seeking to make appointments by agreement are well advised to propose only persons who are on appropriate lists and can be seen to have the necessary knowledge and experience: although even that precaution is not always sufficient.

In *Pratt* v. *Swanmore Builders and Baker (1980)* (15 BLR 37) an arbitrator appointed by the Chartered Institute of Arbitrators who had 'shown himself to be quite incompetent to conduct the arbitration' and had 'allowed the arbitration to be reduced to such a state that there was no prospect of justice being done' was removed by order of the High Court under section 23 of the 1950 Act. Fortunately such occurrences are rare, and especially so where the appointing authority administers a rigorous examination system for aspiring arbitrators.

Lay parties, and regrettably even some appointing authorities, sometimes assume that the qualification *ACIArb* (Associate of the Chartered Institute of Arbitrators) implies competence to conduct arbitrations. In fact that qualification is available after little more than the successful completion of a training weekend. It is, or should be, but the starting point in a study of arbitration procedure and practice extending over several years, such as that administered by the Chartered Institute of Arbitrators, which involves both theoretical and practical training, and culminates in examinations for qualification in the senior grade of *Fellow*. Thereafter, following upon satisfactory completion of a period of pupillage and an interview, a candidate may be admitted to the Institute's panels of arbitrators.

2 Arbitration agreements

SYNOPSIS

The principal requirement of an arbitration agreement, in order that the arbitration may be governed by the 1950 and 1979 Acts, is that it shall be in writing. Hence the inclusion or reference to one of the standard forms of construction contract incorporating an arbitration clause, in a properly executed contract, provides a valid arbitration agreement.

Many necessary terms of an arbitration agreement are deemed by the 1950 Act to have been included unless there is specific provision to the contrary. But arbitration agreements may go much further in prescribing rules for the conduct of the arbitration and in certain circumstances for exclusion of rights of appeal.

DEFINITION

'In this part of the Act, unless the context otherwise requires, the expression "arbitration agreement" means a written agreement to submit present or future differences to arbitration, whether an arbitrator is named therein or not' (1950 Act, section 32).

It follows that there could be difficulty in establishing the existence of a valid arbitration agreement where the contract is an oral one, or where for example a written tender has been accepted orally or impliedly. However, where there is an oral or implied acceptance of a written tender or offer which is expressly based upon a standard form incorporating an arbitration agreement, it is arguable that the standard form, and hence the arbitration agreement, is incorporated in the contract.

AGREEMENTS TO REFER

Arbitration agreements made before a dispute arises, such as those incorporated in the ICE, FIDIC and JCT forms of contract and in the

FCEC form of subcontract are often termed 'agreements to refer' because they provide for the reference to arbitration of any dispute that may arise from the contract.

AD HOC AGREEMENTS

Where a dispute arises from a contract in which there is no arbitration agreement within the meaning of the 1950 Act it is open to the parties to enter into an '*ad hoc*' arbitration agreement in respect of that dispute. In practice, however, it may be difficult to persuade the parties – or for one party to persuade the other – to enter into any form of agreement once a dispute has arisen.

PROTECTION AGAINST COURT PROCEEDINGS

Where a party to a valid arbitration agreement chooses to ignore that agreement and commences proceedings in court it is open to the other party, before delivering a defence or taking any other step in the proceedings, to apply to the court for a stay of the action under section 4 of the 1950 Act. The applicant must satisfy the court that he is ready and willing to proceed in accordance with the arbitration agreement, and the court must also be satisfied that there is no sufficient reason why the matter should not be referred in accordance with the agreement. Although the power of the court to order a stay of the court action is discretionary it will usually be exercised where these conditions are satisfied.

In *Croudace* v. *The London Borough of Lambeth (1986)* (33 BLR 20) the Court of Appeal, in granting a stay under section 4 of the 1950 Act, held that the absence of a dispute at the time when the court action was commenced did not debar the defendant from applying for a stay of that action.

However in *Chatbrown* v. *Alfred McAlpine Construction (Southern) (1986)* (35 BLR 44) the Court of Appeal upheld a decision of the Official Referee refusing to grant a stay in a case in which the Plaintiff had applied for summary judgment under Order 14 of the Rules of the Supreme Court. In that action the Defendant had sought to set off expenses allegedly incurred as a result of delays by the Plaintiff, his subcontractor, against admitted entitlement to interim payment for work done.

Special provision overriding section 4 of the 1950 Act, whereby in certain circumstances a *consumer* may opt out of an arbitration agreement, has been enacted in the *Consumer Arbitration Agreements Act 1988*. That Act defines a consumer as a person who *neither makes the contract in the course of a business nor holds himself out as doing so*: where the *other party makes the contract in the course of a business*: and requires that the goods supplied under the contract must be *of a type ordinarily supplied for private use or consumption*. The definition expressly excludes sales by auction or by competitive tender. The Act is, because of these limitations, unlikely to be relevant to construction contracts.

CONSTITUTION OF TRIBUNAL

Generally, and almost invariably in domestic construction contracts, the reference is to a single arbitrator. Such an intention need not, however, be defined in the arbitration agreement, because under section 6 of the 1950 Act a provision to this effect is deemed to be included in every arbitration agreement unless some other mode of reference is provided.

APPOINTMENT PROCEDURE

Although it is desirable that the parties to an arbitration agreement should define the way in which the arbitrator (if not named in the agreement) is to be appointed, especially if it is their intention that he should be appropriately qualified to deal with the technicalities involved, failure to include such a definition is not fatal to the operation of the arbitration agreement. Section 10 of the 1950 Act provides a procedure for the appointment of an arbitrator by the High Court where the parties are unable to agree upon an appointment, and it also deals with vacancies that arise where the appointed arbitrator is incapable of acting, or refuses to act, or dies.

Until the 1979 Act became law there remained a situation in which deadlock could be reached because of a breakdown in the machinery for the appointment of the arbitrator. There was no provision under the 1950 Act for filling a vacancy where the appointing authority named in the arbitration agreement refused, or failed within the time specified, to make an appointment. This defect has been remedied under section 6 of the 1979 Act, which amends section 10 of the 1950 Act to enable the High Court to make an appointment in this situation.

ARBITRATION RULES

It is open to the parties to include in either form of an arbitration agreement, or in an addendum thereto made either before or after a dispute has arisen, any rules for the conduct of the arbitration, provided that such rules are not unlawful or contrary to public policy. Standard rules are published by such bodies as the Chartered Institute of Arbitrators, the London Court of International Arbitration, the United Nations Commission on International Trade Law (the 'UNICITRAL Rules'), the International Chamber of Commerce, the Institution of Civil Engineers, and others, usually with the objectives of defining time limits for the various stages of the proceedings and of strengthening the arbitrator's powers to deal with dilatory or obstructive parties.

Provision is also made under section 5 of the 1979 Act for applications either by the arbitrator or by one of the parties to the reference for extending the powers of the arbitrator to deal with disobedient parties. Under this section the arbitrator may be empowered to proceed with the reference in default of appearance or of any other act in the manner in which a judge of the High Court might proceed where a party fails to comply with an order of the court or a requirement of the rules of court.

In some references the parties voluntarily agree, at the beginning of or during the proceedings, that the arbitrator shall have the powers referred to in that section.

Similarly the parties may adopt by agreement any lawful rules they themselves formulate. Matters that may well be covered in such rules, especially where the dispute involves a small sum or where the issues arising from it are of a technical rather than a legal nature, are limitation of the right to appoint counsel or solicitors as the parties' representatives, limitations as to the appointment of technical experts, and provision for the dispute to be determined on written evidence only, perhaps in conjunction with an inspection of the works in question.

Again, recognizing that in a major dispute time spent at the hearing is expensive in that both parties' counsel, solicitors, experts and witnesses of fact, in addition to the arbitrator, are present, economy is achieved by reducing that time to a minimum. An agreement that the opening address of the claimant's counsel, and where practicable, that of the respondent's counsel, should be put into writing and submitted to the arbitrator a few days before the hearing begins enables him to study those addresses in advance of the hearing, and avoids the unproductive tedium of long addresses read at dictation speed by counsel, often from prepared scripts, and laboriously written down by the arbitrator. The same principle may be applied to the evidence-in-chief of experts and of witnesses of fact, so that the substance of the hearing begins with cross-examination.

An even greater advantage derives from the submission in writing of counsel's closing addresses. For at the conclusion of a hearing counsel usually require time to consider the evidence and to formulate their submissions; and that time is not available if they have to rise immediately after the conclusion of the evidence. By adopting a rule that the respondent's closing address shall be submitted in writing within say seven days of the close of the hearing, and the claimant's address within a further seven days thereafter, both counsel have time to prepare their submissions, and time is not wasted by all of those who attend the hearing.

Much benefit can often be gained by the parties from an agreement to adopt rules appropriate to the magnitude and nature of the dispute, as a means of minimizing costs and expediting the resolution of the dispute. It may however be difficult to foresee, at the time of making an agreement to refer, the nature and magnitude of disputes likely to arise: if they were foreseeable they would be dealt with in the terms of the contract. Hence the parties may find a need to amend or supplement their arbitration agreement after a dispute has arisen, and at this stage the relationship between the parties is often not conducive to agreement of any kind. Nevertheless, a party who finds an opponent obstructive to sensible rules of procedure should ensure that the proposals are clearly set out in a letter to the other party and brought to the notice of the arbitrator, with a request that they are taken into consideration in the arbitrator's award of costs.

Where it is agreed between the parties that a published set of rules shall apply it is important that there should be a definition of the relevant edition of such rules, in order to avoid the possibility of contention, and consequent delay. Such definition could either state the publication date of the edition that is to apply, or could specify that the relevant edition shall be that current at the date of the arbitration agreement or at the date of commencement of the arbitration.

EXCLUSION AGREEMENTS

The 1979 Act, which introduces a new procedure for judicial review of points of law arising either from an award or during the course of a reference, also provides a limited right of the parties to exclude such review. An agreement for this purpose is appropriately termed an 'exclusion agreement' (see SD/14), and in domestic contracts it is valid only if entered into after the commencement of the arbitration.

THE ICE CONDITIONS OF CONTRACT

The current (sixth) edition of the ICE Conditions of Contract, and earlier editions of the form, incorporates as Clause 66 an arbitration agreement within the meaning of the 1950 Act. Hence, provided that the form is signed by both parties or is referred to in a properly constituted contract (that is, one incorporating a valid offer and a valid acceptance), it will bring the arbitration within the scope of the 1950 Act. The clause also makes provision for the arbitrator to be appointed, failing agreement between the parties, by the President (or a Vice-President) of the Institution of Civil Engineers; and for the President to make, on the application of either party, a further appointment to fill any vacancy created by the nominee refusing to act, by his removal by the court, by incapacity or by his dying, where the parties do not themselves fill the vacancy within one month.

The Institution of Civil Engineers publishes a *List of Arbitrators* wherein are published the names and brief curriculum vitae of all of those persons who have qualified with the Institution as arbitrators. Although not bound to do so it is to be expected that the parties to a dispute, when attempting to agree on the appointment of an arbitrator, or the President of the ICE when making an appointment, will choose a person whose name appears in that list. By so doing the parties, or the President, are able to ensure that the person chosen is properly qualified as an arbitrator and has knowledge and experience relevant to the matters in dispute.

Earlier editions of the ICE Conditions made provision for arbitrations to be conducted, if the parties so agreed or if the President of the ICE so directed, in accordance with the Institution of Civil Engineers' Arbitration Procedure (1973). Under the June 1985 revision of the fifth edition adoption of the ICE Arbitration Procedure (1983) became mandatory, and that provision remains applicable to the sixth edition.

Both the current and earlier editions of the ICE Conditions make provision for the reference of any dispute or difference that may arise to the engineer for his formal decision under Clause 66, as a prerequisite to the aggrieved party's giving notice requiring the dispute to be referred to arbitration. The engineer is then allowed a period of three months (or, where a Certificate of Completion has not been issued, one month) in which to give his Clause 66 decision. If that decision is not given within the prescribed time, or if it is unacceptable to either party, then that party must give a written *Notice to Refer* within three months of the date of the decision, or of the latest date by which the decision should have been given. Failure to give such notice results in the engineer's Clause 66 decision becoming final and binding on both parties.

One of the objections to that requirement is that it adds a further three months (or one month) to the overall time taken to resolve the dispute; for the likelihood that the engineer's decision under Clause 66 will be any more acceptable to the aggrieved party than his earlier decision from which the dispute arose is remote indeed. Furthermore, if it is thought that the need to make a formal decision will give warning to the engineer that his actions may lead to arbitration, then removal of the requirement should concentrate the engineer's mind on the seriousness of the position at an earlier stage in the negotiations. For, in civil engineering contract disputes, as in any other type of dispute, the parties rarely invoke arbitration or litigation without having exhausted their attempts to negotiate a settlement.

A further objection to what has become known as the 'two-stage' procedure under the ICE Conditions is that it may lead to attempts by an unscrupulous party to argue that a decision given more than three months earlier by the engineer constituted his decision under Clause 66, and that as the aggrieved party did not within the prescribed period give notice of arbitration he is now deemed to have accepted the engineer's decision.

However, the Court of Appeal decided, in *Monmouthshire CC* v. *Costelloe*

& Kemple (1965) (5 BLR 83), that an engineer must make it quite clear that a statement is intended to be a decision under Clause 66: and furthermore that a decision under that clause cannot be given by the engineer until the aggrieved party has specifically requested an engineer's decision under Clause 66. Thus it is the aggrieved party, and not the engineer, who is entitled to initiate a reference under Clause 66.

Earlier editions of the ICE Conditions sought to preclude the commencement of arbitration proceedings until after completion, or alleged completion, of the works. Clearly the laudable aims of this requirement were to avoid the likely disruption to progress of the work where those involved in construction might find their time and energy diverted to the less productive actitivies involved in the reference and to avoid unnecessary multiplication of references to arbitration. However, an absolute prohibition of such proceedings during the course of construction could prevent a contractor from obtaining the payments needed to continue in business, thereby making it impossible for him to complete the works. Accordingly the somewhat Draconian requirements of earlier editions of the conditions were, under the fifth edition, softened by providing for immediate arbitration of any dispute arising from Clause 12 of the conditions (which clause relates to payments in respect of adverse physical conditions and artificial obstructions) or from the withholding by the engineer of any certificate.

It was held in *Farr (A. E.)* v. *Ministry of Transport (1960)* (1 WLR 956) that failure by the engineer to certify more than £15 000 of a proper claim by the contractor of £20 000 constituted 'withholding a certificate', in that case because of an error of law by the engineer; and accordingly the contractor was entitled to immediate arbitration. Whether the same rule would have applied to a reduction in the value of the certificate resulting from measurement differences is not clear. If the rule does so apply then effectively the contractor has a right to immediate arbitration of any substantial difference arising from the contract, but that right would not extend to cover a difference arising from an allegation of breach of contract.

Sixth edition of the ICE Conditions of Contract

Important changes to Clause 66 of the ICE Conditions were first introduced in June 1985 as amendments to the fifth edition: most of which amendments were retained without substantial alteration in the sixth edition, and are as follows:

(i) the removal of any bar to the commencement of arbitrations before completion of the works
(ii) a reduction to one month of the time allowed for the engineer's decision in cases where the arbitration is begun before completion of the works and

(iii) mandatory adoption of the Institution of Civil Engineers' Arbitration Procedure (1983).

Additionally, provision has been made for a further intermediate stage of dispute resolution, namely conciliation (see below); and the clause has been rearranged into a simpler and clearer format.

The first major change probably results from recognition of the impracticability of imposing any bar to 'immediate' arbitration in the light of the *Farr* decision (above). Nevertheless it is submitted that the parties should not abuse their newly gained freedom by requiring immediate arbitration unnecessarily. One purpose of the original rule was to obviate the need for a multiplicity of arbitrations arising from a single contract, in order to minimize costs. An arbitrator who finds that costs have been incurred unnecessarily by premature references of disputes, where they should have been dealt with as a single reference on completion, might well take this fact into consideration in his award of costs.

The second change probably originates from a laudable desire to ensure that disputes arising during construction are settled as promptly as possible, because of their likely disruptive effect on progress.

The third change imposes upon the parties the obligation to comply with the ICE Arbitration Procedure (1983), which Procedure is discussed below.

Conciliation

The sixth edition of the ICE Conditions includes a provision, first introduced by the ICE in the *Minor Works* form (see below), for conciliation under the *ICE Conciliation Procedure 1988*. This provision may be invoked by either party after the engineer has given his Clause 66 decision or the time for giving such decision has expired, but before the dispute is referred to arbitration.

The procedure differs from the usual negotiations between the contractor and the engineer in attempting to settle differences by agreement in that it provides for the appointment, either by agreement or upon application to the President of the ICE, of a conciliator, whose function is to make a recommendation. He does so after receipt of written submissions from the parties: and he may, having given notice to the parties, visit and inspect the site or the subject matter in dispute. He may also convene meetings with both parties together, or separate meetings with either party, and may take evidence and hear submissions; but he is not bound by the rules of evidence. The conciliator is required to conclude the conciliation as soon as possible, and in any event within two months of his appointment unless the parties otherwise agree. The procedure also provides that the conciliator shall not be appointed arbitrator in any subsequent arbitration whether arising out of the dispute or otherwise, unless the parties otherwise agree.

The conciliator's recommendation is of course only a recommendation: it is not binding on the parties. Hence the conciliation procedure is effectively an attempt to formalize negotiations between the parties which almost invariably precede, and usually continue during, arbitration proceedings. For that reason it is sometimes argued that the procedure is otiose: the parties are, and always have been, free to negotiate in whatever way they please, whether or not the ICE conciliation procedure, or any other formalized negotiation procedure, applies.

However, there is one very good reason for the introduction of the procedure. In many disputes the parties, while professing that their cases are invincible, would privately welcome a negotiated settlement of their differences. To suggest an informal *without prejudice* meeting would however be seen as implying a weakness in their case which would be detrimental if the dispute has to run its full course to and through the arbitration hearing. Hence each party would prefer that its opponent takes the initiative in proposing a meeting. It is, in many cases, only when it becomes clear that time for such an initiative has run out that a settlement is achieved: either on the doorstep of the arbitration court or, in some cases, while the arbitrator sits waiting for the parties to indicate that they are ready to proceed, or during the course of the hearing.

If the imposition of a formalized conciliation procedure has the effect of overcoming the parties' reluctance to propose some such procedure, then it may well result in earlier settlements with substantial savings in cost, and may facilitate settlements that might not otherwise be achieved.

The ICE Conditions of Contract for Minor Works

This form, first published in 1988, includes as Clause 11 a procedure for the resolution of disputes, under which reference to the engineer, a mandatory requirement of all of the main forms of contract published by the ICE, is omitted. The clause requires firstly that a *Notice of Dispute* be served by one party on the other; after which either party may, within 28 days, serve on the other a notice requiring the dispute to be considered under the ICE Conciliation Procedure. Alternatively the *Notice of Dispute* may, within 28 days, be followed by a *Notice to Refer*.

In the event that neither such notice is served within the 28-day period the *Notice of Dispute* is deemed to have been withdrawn. The Minor Works form also provides for mandatory adoption of the ICE Arbitration Procedure (1983) and for the use of the Short Procedure in Part F of that Procedure.

The Short Procedure includes *inter alia* a provision, under Rule 21.1, that, unless the parties agree to the contrary, the arbitrator is deprived of his power to award costs of the reference, and costs of the award are to be borne in equal shares by both parties. That rule is however in conflict with, and is rendered void by, section 18(3) of the 1950 Act, which sensibly provides that such an agreement is valid only if it is entered into *after a dispute has arisen*.

It is submitted that the parties should carefully consider, before making a valid agreement (i.e. an agreement made after the dispute has arisen) to adopt Rule 21.1, what benefit is to be gained from that rule. Under it, the arbitrator is deprived of one of his most valuable powers to control procedural matters and to ensure that additional costs occasioned by dilatory or obstructive behaviour by either party are borne by that party.

THE INSTITUTION OF CIVIL ENGINEERS' ARBITRATION PROCEDURE (1983)

Unlike its predecessor, the 1983 version of the ICE Procedure is not a booklet explaining arbitration procedure, but is a leaflet containing a set of arbitration rules. Separate versions are issued for England and Wales, and for Scotland. Many of the rules merely draw attention to powers which are already available under the two main statutes or at common law, but in addition there are several rules that obviate the need for applications to the High Court, such as Rule 6.1 Protective Measures: Security for Costs, etc.: for which applications to the High Court under section 12 of the 1950 Act would otherwise be necessary.

The power to consolidate references relating to the same subject-matter, available in litigation by use of a third-party notice, is not generally available in arbitration. Rule 7.1 of the ICE Arbitration Procedure provides for such consolidation by order of the arbitrator, *where the several disputes have been referred to the same arbitrator*, either by agreement of all of the parties concerned or upon the application of one of the parties being a party to all of the contracts involved.

The power to consolidate by agreement of all of the parties is of course available whether or not Rule 7.1 applies. Conversely, where all of the parties do not agree thereto, the arbitrator's power to order consolidation is, it is submitted, conditional upon the parties who do not agree being bound by the ICE Arbitration Procedure. In the usual employer/main contractor/subcontractor type of multi-party dispute it is to be expected that the party seeking consolidation will be the main contractor, who will be a party to both the main contract and the subcontract, and will therefore be able to invoke the provisions of Rule 7.1.

However, what may perhaps be more generally effective in such cases is not the arbitrator's power to order consolidation, but the use of his discretion in his award of costs where it appears that economy could be achieved by concurrent hearings. An indication by the arbitrator to parties who object to consolidation that he may, in the exercise of his discretion as to the award of costs, take account of an unreasonable objection, may result in such parties agreeing to the more efficient procedure.

Probably the most important of the powers included in the rules is that under Rule 14 to make summary awards. Regrettably that rule is ambiguous in that the phrase 'at any time' in the first line may or may not be qualified by the closing words '. . . after determination of all the issues in the arbitration . . .'. If the phrase is so qualified then the power is available without any such rules: if it is not, then the power is, it is submitted, so wide as to be open to challenge on the ground of misconduct.

Rule 21.1 of the Short Procedure provides that, unless the parties otherwise agree, the arbitrator has no power to award costs to either party, and the arbitrator's own charges are to be paid in equal shares by the parties.

The first part of this provision is presumably based on an intention to discourage the appointment of legal or technical representatives where the matters in dispute are of a minor nature: and the second part follows logically from the first. However, for reasons given above, it is submitted that the parties should consider carefully the full implications of the rule before agreeing, after the dispute has arisen, to its inclusion.

Under Rule 24.7 the arbitrator in an *Interim Arbitration* (defined in Rule 24.1 as an arbitration that is to proceed before completion or alleged completion of the works) is empowered to direct that Part F (Short Procedure) and/or Part G (Special Procedure for Experts) shall apply to the Interim Arbitration. Part F includes Rule 21.1 referred to above: and it is submitted that an arbitrator's direction that that rule shall apply would be void under section 18(3) of the 1950 Act, unless the parties voluntarily agree to adopt that rule *after the dispute has arisen.*

THE FCEC FORM OF SUBCONTRACT

The Form of Subcontract published by the Federation of Civil Engineering Contractors, which is designed for use in conjunction with the ICE Conditions of Contract, includes as Clause 18(1) a simple but adequate arbitration agreement. This provides for the arbitrator to be appointed by agreement between the parties, or failing agreement by the President of the ICE.

A situation that often arises in work performed by a subcontractor is that the real dispute is between the employer and the subcontractor: the main contractor, although a party to both main and subcontracts, is concerned only that his obligation under one of the contracts is reflected in a corresponding right under the other. If the subcontractor claims additional payment the main contractor will be willing to make that payment provided that he can recover the cost from the employer; and similarly if the employer rejects any parts of the works then the main contractor will be concerned to see that the rejection may be passed on as an obligation upon the subcontractor to make good the defects.

Clause 18(2) of the FCEC Form deals with this situation as far as it is possible to do so, by requiring the subcontractor to accept the arbitrator appointed under the main contract to act also under the subcontract. Hence where the dispute orginates as a claim by the employer (for example, an allegation of defective work) it is usually possible for the main contractor to use this subclause to ensure that the two references are referred to the same arbitrator.

Where the dispute originates as a claim by the subcontractor consolidation of a corresponding dispute between main contractor and employer can only be effected by agreement. It would, however, be possible for the main contractor to initiate an arbitration against the employer in respect of his subcontractor's claim, and then to require the subcontractor to accept the same arbitrator, under Clause 18(2).

The co-existence of Clauses 18(1) and 18(2) of the FCEC Form sometimes gives rise to a dispute as to the validity of an appointment under the former provision. Where the subcontractor gives notice under Clause 18(1) requiring a dispute to be referred to arbitration, the main contractor may, under the September 1984 revision of the FCEC Form, invoke Clause 18(2), giving notice in writing requiring the dispute to be dealt with *jointly with the dispute under the Main Contract in accordance with the provisions of Clause 66 thereof.*

Clearly such a notice is valid only if a dispute has in fact arisen under the main contract; and that raises the question of how the existence of such a dispute is to be defined. It is submitted that an engineer's rejection of, or failure to agree to, a claim under the main contract does not in itself constitute a dispute; and that a dispute within the meaning of Clause 18(2) impliedly arises only where the main contractor invokes Clause 66 of the main contract.

Clause 18(3) of the FCEC Form covers the less likely possibility of the main contractor being involved in court proceedings in respect of a dispute that concerns the subcontract works. In such a situation the main contractor would wish to serve a third-party notice upon the subcontractor, but could be prevented from proceeding in such a manner by the subcontractor's applying for a stay of the court proceedings under section 4 of the 1950 Act, on the ground that there is a valid arbitration agreement between the parties, to which the dispute should be referred. Clause 18(3) empowers the main contractor in this situation to abrogate the arbitration agreement.

THE FIDIC CONDITIONS OF CONTRACT

Under the Conditions of Contract (International) for Works of Civil Engineering Construction, better known as the FIDIC Conditions, the two-stage procedure of the ICE Conditions applies, with minor modifications to the wording. The arbitration agreement includes a provision that the Rules of Conciliation and Arbitration of the International Chamber of Commerce (ICC) shall apply.

Those rules provide *inter alia* for arbitrators', and where appropriate administering authorities', fees to be based upon a percentage of the sum in dispute, and for the tribunal to be of three, with a lawyer as chairman. Quite apart from the undesirability of having a legal person to determine questions which are predominantly technical, the costs of arbitration in this form is likely to be excessive.

Consideration is currently being given by the ICE to urging the adoption of an alternative arbitration clause, specifying arbitration under the Rules of the London Court of International Arbitration or other rules based upon English arbitration procedure. It is to be hoped that that change will be adopted by the various bodies concerned in drafting the FIDIC Conditions: meanwhile parties entering into contracts based upon the rules are well advised to amend the arbitration clause to provide for London arbitration.

THE STANDARD FORM OF BUILDING CONTRACT

The Standard Form of Building Contract, published by the Joint Contracts Tribunal and often referred to as the JCT Form, or incorrectly as the RIBA Form, includes an arbitration agreement as Article 5 of the Form of Agreement. This provides for appointment of the arbitrator by agreement, or failing agreement within 14 days of one party serving upon the other a written notice to concur in the appointment of an arbitrator, for the arbitrator to be appointed by the President or a Vice-President of the RIBA.

The 1980 editions of the JCT Form introduce a new requirement corresponding to that of the FCEC Form, which seeks to ensure that where a dispute between the employer and the contractor is concerned with issues that are substantially the same as those between the contractor and a nominated subcontractor, both references to arbitration are dealt with by the same arbitrator. The Form does, however, recognize that this sensible provision is not always practicable, because the qualifications required of the arbitrator may not be the same in both cases.

Amendment 6 to the JCT Form, introduced in July 1988, incorporates a number of improvements to the earlier version. Clause 41.1 requires the initiating party to give written notice that a dispute be referred to arbitration: while Clause 41.1 specifically empowers the arbitrator to rectify the contract. Provision is made, under Clause 41.8, for the appointing authority to re-appoint where a vacancy arises from the arbitrator dying or otherwise ceasing to act.

Perhaps the most significant change, however, is the introduction of the JCT Arbitration Rules, dated 18 July 1988, which become mandatory under a new Clause 41.9 of the JCT Form. The rules, which are referred to in many of the standard forms of contract published by the JCT, impose strict time limits on the various stages in an arbitration, and encourage the use of shortened procedures such as the elimination of a hearing.

3 Powers of an arbitrator

SYNOPSIS

An arbitrator derives his powers from the arbitration agreement between the parties (see Chapter 2) and from the 1950 and 1979 Acts. Provided that the agreement is valid and that he has been properly appointed the two Acts give him the basic powers needed to conduct the proceedings and to make an enforceable award. His powers may be extended by agreement between the parties, and where he considers it necessary for the proper conduct of the reference the arbitrator may seek such extension of his powers from the parties, or from the High Court.

IRREVOCABILITY OF AUTHORITY

Section 1 of the 1950 Act provides that the arbitrator's authority is irrevocable except by leave of the High Court or a judge thereof. Although the parties are in general able to adopt by agreement rules to expedite procedures, and are able to agree as to the appointment of the arbitrator, the Act sensibly excludes from these rights the right to remove an arbitrator once he has been appointed, except by leave of the High Court; and that court would, implicitly, have to be satisfied that there were adequate grounds for his removal.

The appointed arbitrator may, however, refuse to act, or become incapable of acting, or die. Section 10(b) of the 1950 Act provides for all of these eventualities by empowering the High Court, where the arbitration agreement is silent, to make the necessary re-appointment. Many of the standard forms of construction contract, including the ICE Conditions, make provision for the President of the relevant professional body to re-appoint when a vacancy arises from one of these clauses.

CONDUCT OF PROCEEDINGS

The principal power required by the arbitrator, namely that to order the parties to appear before him and to produce all relevant documents, is conferred under section 12 of the 1950 Act by way of a deemed term in arbitration agreements that the parties shall do these things, and 'all other things which during the proceedings on the reference the arbitrator or umpire may require'. Certain other powers that may be needed, such as the power to subpoena a witness to appear and to give evidence at the hearing, are available to the parties but only upon application to the High Court.

Section 12(3) of the 1950 Act empowers the arbitrator to administer oaths or to take affirmations. Usually the arbitrator will require witnesses to swear or to affirm, and he should always do so where credibility is in question.

POWERS RELATING TO THE AWARD

Other important powers conferred to the 1950 Act are:

Under section 13: Power to make an award at any time: subject to the power of the High Court, on the application of either party, to remove an arbitrator who has failed to use reasonable dispatch in the proceedings.
Under section 14: Power to make an interim award (see Chapter 8).
Under section 15: Power to order specific performance (see Chapter 8).
Under section 17: Power to correct mistakes or errors (see Chapter 10).
Under section 18: Power to award costs (see Chapter 9).
Under section 19A: Power to award interest (see Chapter 8).

POWER TO DEAL WITH A RELUCTANT PARTY

It sometimes happens before an arbitration is begun, or during its preliminary stages, that a party becomes aware of the weakness of its case and seeks to delay the proceedings as long as possible, in order to defer the day of settlement. The strategy usually adopted in such cases is to ignore the arbitrator's orders or to apply for excessive periods of time for preparation of pleadings; to apply for extensions; to apply for 'further and better particulars' (see Chapter 5) in unnecessary detail, or to fail to appear at meetings or at the hearing.

The practice in the past, and indeed that recommended in earlier editions of this book, has been for the arbitrator to discourage such tactics in a gentle way, taking the greatest care not to make any order that might be reversed in the courts. More recently it has become recognized that the courts will not interfere with the arbitrator's conduct of procedural matters except where there is an infringement of the rules of natural justice or where an arbitrator is seen to be biased (*Turner (East Asia) Pte* v. *Builders*

Federal (Hong Kong) and Josef Gartner & Co.) (1988) (42 BLR 122). Arbitrators are encouraged to deal robustly with parties seeking to cause delay: either by ensuring that costs occasioned by delay are borne by the defaulting party; or, in appropriate circumstances, by proceeding with the reference in the absence of a pleading or, at a hearing, in the absence of a party.

Where provision is made, for example at a preliminary meeting, for the likelihood that delays may occur, the inevitable result is that delays *will* occur: the party seeking to cause delay being aware that it is unlikely to incur a penalty for failing to comply with the arbitrator's directions. It is suggested therefore that no such provision should be made. The timetable for the various stages of the preliminaries and for the hearing should wherever possible be determined at the preliminary meeting (see Chapter 5) and the parties should be warned that costs occasioned by delay will be borne by the party responsible for the delay *in any event*.

Where the respondent fails to serve his defence by the due date or to apply for an extension of time it is not unreasonable for the arbitrator to infer that that party has no defence. In such a case, after due warning has been given, the arbitrator should convene a hearing, at which the respondent is permitted to challenge the claimant's evidence but not to make any other submission, since it cannot have been pleaded.

Where a party fails to respond to or to comply with the arbitrator's directions, or indicates that it does not intend to be present or to be represented at a hearing, the arbitrator should issue a formal notice of the hearing marked 'PEREMPTORY', and should state in that notice his intention, if necessary, and on the application of the party in attendance, to proceed *ex parte* to hear the evidence of that party and thereafter to make his award.

Where a party unexpectedly fails to appear, either in person or by representation, at a hearing, the arbitrator may suggest to the party present that it tries to contact the party in default by telephone, to ascertain if the party has been delayed *en route* or if there is some other explanation for its absence. If and when it becomes clear that no useful purpose will be served by waiting, the arbitrator must adjourn the hearing to a later date, and must serve a peremptory notice, as above, of the adjourned hearing.

In either situation an arbitrator conducting an *ex parte* hearing has a duty to ensure that the claimant's case is proved before making an award in his favour. Where the arbitrator finds reason to doubt the validity of the claim or of its evaluation he *must* ensure that his doubts are expressed at the hearing, in order that the party in attendance has an opportunity to deal with and to challenge those doubts in the same way that it would have dealt with or challenged the evidence or submissions of the other party had it been present (*Fisher and Another* v. *P. G. Wellfair (1981)* (19 BLR 52)).

Until the 1979 Act became law a problem could arise where a reluctant respondent ignored the arbitrator's orders as to interlocutory matters such as pleadings and discovery, but arrived at the hearing in order to present a defence. The arbitrator had no power to refuse permission to the respondent to take part: but to allow the hearing to proceed would put the claimant at a disadvantage because he had not been forewarned of the respondent's defence. So he would have to grant a further adjournment, having ascertained the nature of the defence, in order that the claimant could deal with the points raised in that defence.

Section 5 of the 1979 Act greatly strengthens the arbitrator's power to deal with such tactics, by permitting him, or the other party, to apply to the High Court for an order extending the arbitrator's powers in a case where a party fails within the time specified, or within a reasonable time, to comply with his orders. An order granted under this section could, subject to any limitation that may be imposed by the court, empower the arbitrator to continue with the reference in the way that a High Court judge might deal with a similar situation in that court: that is, by making an order debarring the respondent from defending the claim at all, leaving it to the claimant to prove his case unhindered by the respondent. Such an order could where appropriate be applied to that part of the claim which is affected by the respondent's failure to comply with the arbitrator's order.

Similarly, where the claimant fails to proceed with reasonable dispatch the arbitrator or the respondent may now apply to the court for an order enabling the arbitrator to strike out all or the relevant parts of a claim.

THE RELUCTANT CLAIMANT

Although it is less likely that delays will be caused by the claimant than by the respondent, that situation does sometimes arise: for example, where the claimant is confronted with a counterclaim which may equal or exceed the amount of the claim: where unforeseen difficulty is encountered by the claimant or its representatives in preparing its case: or where doubts as to the validity of the claimant's case increase during the interlocutory proceedings. In the former situation it sometimes happens that the respondent takes over from the claimant in seeking to urge expedition; when the techniques referred to above need to be applied to the reluctant claimant.

Where neither party seeks to bring the dispute to a hearing the arbitrator who, rightly, considers it to be part of his duty to do so, often encounters difficulty. The parties may, for example, agree between themselves to extensions of time for delivery of pleadings, and then advise the arbitrator of their agreement. Whether or not the arbitrator should seek to interfere in matters agreed between the parties is debatable: but it is submitted that he should at least ensure that the parties themselves – as distinct from their representatives – are aware of the causes of the delay.

In *Bremer Vulkan Schiffbau und Maschinenfabrik* v. *South India Shipping Corporation (1981)* (2 WLR 141) it was affirmed in the House of Lords that an arbitrator did not have the power available to a judge to dismiss a claim for want of prosecution. That decision, which has been widely criticized, has been the subject of a statutory reversal under section 102 of the Courts and Legal Services Act 1990: which section became effective on 1 January 1992 under the C&LSA 1990 (Commencement No. 7) Order (SI 1991 2930).

Section 102 empowers an arbitrator to make an award dismissing a claim referred to him provided that (a) there has been inordinate and inexcusable delay on the part of the claimant in pursuing the claim and (b) that the delay will give rise to a substantial risk that it is not possible to have a fair resolution of the issues in the claim, or that it has caused, or is likely to cause, serious prejudice to the respondent.

POWER TO REVISE ENGINEER'S DECISIONS

Under Clause 66 of the ICE Conditions an arbitrator appointed to determine disputes arising from a contract under that form has '. . . full power to open up review and revise any decision, opinion, instruction, direction, certificate or valuation of the engineer . . .'. Similar provision is made in Article 5.3 of the JCT Standard Form of Building Contract empowering an arbitrator appointed under that form '. . . to open up, review and revise any certificate, opinion, decision, requirement or notice and to determine all matters in dispute . . .'. It was held by the Court of Appeal in *Northern Regional Health Authority* v. *Derek Crouch Construction (1984)* (26 BLR 1), in an action arising from a contract on the JCT Standard Form of Building Contract, that the wide powers given to an arbitrator to review the architect's exercise of his discretionary powers under the contract were not available to the court. Similarly the court would have no power to review the engineer's exercise of his discretion under the ICE Conditions of Contract.

Under section 100 of the Courts and Legal Services Act 1990, which section came into force on 1 April 1991, a new section 43A of the Supreme Court Act 1981 was inserted, as follows:

43A. *In any cause or matter proceeding in the High Court in connection with any contract incorporating an arbitration agreement which confers specific powers upon the arbitrator, the High Court may, if all parties to the agreement agree, exercise any such powers.*

Although that statute appears to reverse the *Crouch* decision, it does so only with the agreement of all of the parties to the arbitration agreement: which agreement, it is submitted, would not generally be forthcoming. A party in whose favour the engineer's decision under Clause 66 of the ICE Conditions, or the architect's decision under Article 5.3 of the JCT Form, is given, would be unlikely to agree to a power being conferred on the court to overturn such a decision.

However, where an official referee is appointed as *arbitrator*, under section 11 of the 1950 Act, as amended, he would of course be able to exercise all of the powers conferred on the arbitrator by the agreement.

4 Appointment of the arbitrator

SYNOPSIS

A party wishing to refer a dispute to arbitration should comply strictly with the terms of the arbitration agreement in giving notice of arbitration and in initiating the appointment of the arbitrator. The person appointed must be properly qualified to act. Provision is made in the two main statutes for filling a vacancy that arises through the breakdown of machinery for appointment or through the appointed person ceasing to be available.

PROCEDURE FOR APPOINTMENT

The first rule in initiating proceedings is to comply strictly with the arbitration agreement. Thus for instance if the contract incorporates the ICE Conditions of Contract or some derivative of that document such as FIDIC it is necessary to give notice to the engineer that a dispute has arisen, specifying the cause of the dispute, and requiring his decision under Clause 66 of the contract. If he replies within three months* giving an unsatisfactory decision or if he fails to reply within three months* then in either case the arbitration proper may be commenced.

The ICE and many other forms of contract provide for the appointment of an arbitrator *failing agreement* by the named authority: the President of the ICE in the case of a contract under that Institution's contract. Hence an attempt should be made to agree upon an arbitrator, the usual form of such an attempt being a letter from the applicant putting forward the names of up to three engineers, any one of whom would be acceptable to the applicant (see SD/7). Where the ICE Conditions apply the Notice of Arbitration and Notice to Concur may be given on Form ArbICE published by the Institution (see SD/4 and 5).

* One month in the case of a notice given before completion or alleged completion of the works, where the sixth edition of the ICE Conditions apply, or where a contract based on the fifth edition is subject to the June 1985 amendment. See p. 14.

If the other party – usually the respondent – disagrees with all of the suggestions, or if he fails to reply within a reasonable period (one calendar month under the ICE form) then application may be made to the President of the ICE (see SD/5) or in the case of other forms of agreement to the appointing authority named in the agreement. The ICE and many other authorities charge a fee for the administrative work associated with the appointment: in the case of the ICE the current fee is £80 plus VAT.

QUALIFICATIONS OF THE ARBITRATOR

The first and most important qualification of any arbitrator is that he must not have any interest in or relationship with either party such as might impair his impartiality. He must not be a friend or relative of either party; nor must he have any prior knowledge of the subject matter of the dispute, as distinct from a general knowledge of the type of work involved in the dispute, which is necessary. He must not have a financial interest in either party, for example as a shareholder or as a consultant. A prior relationship that no longer exists or a distant relationship by blood or by marriage need not be a bar provided that the arbitrator knows that it will not affect his judgment; but in such a case he should disclose the relationship to both parties and be willing to stand down if either party makes a reasonable objection to his appointment. However the arbitrator should not entertain an unreasonable application for his replacement by another arbitrator if he is satisfied that the real objective is to delay the reference. In such cases he may, having affirmed that his impartiality is unaffected by the former relationship, indicate that he intends to proceed with the reference. But he should ensure that the party raising the objection has an opportunity to apply to the court, for example under section 1 of the 1950 Act, for revocation of his authority. The test for impartiality in such cases is whether or not a reasonable person might consider there to be a risk that the arbitrator is not impartial.

It sometimes happens that the existence of a link with one of the parties becomes known to the arbitrator only after he has accepted the appointment. In such a case he should, if the relationship is such as to affect his impartiality, withdraw as soon as the facts become known to him. In other cases, where his impartiality is not affected, he should nevertheless disclose the relationship to both parties and be prepared to withdraw if either party makes a reasonable request that he do so. He is not debarred from withdrawing by section 1 of the 1950 Act because his action would be to refuse to act: and the parties are free under section 10(b) of that Act to agree upon a replacement or failing agreement to apply to the court for an appointment.

Secondly, the arbitrator should have a general knowledge of the technicalities of the matters in dispute. The extent of this knowledge must depend to a large extent upon the technicality and variety of the issues that arise. It would not, for example, be reasonable to expect the arbitrator to have a specialized knowledge of a wide variety of engineering subjects; in such cases the arbitrator's general knowledge of the variety of subjects might have to be supplemented by expert evidence brought by the parties. On the other hand it is to be expected that a dispute involving road and bridge works would be referred to an arbitrator having that type of experience: not for example to a specialist in water supply.

If the arbitrator finds, either before or after his appointment, that the dispute concerns a matter of which he has prior knowledge – for example, one in which he has given specialist advice, or has made a study for some other purpose, then he should decline or withdraw from the appointment, as the case may be. This is because of the danger that that prior knowledge may influence his decision: his duty in the arbitration is to decide the issues upon the evidence presented.

Thirdly, the arbitrator should have judicial capacity. He should be able to consider and weigh the evidence presented by the parties; to reach a logical decision based upon that evidence upon matters of fact; and to make a just award. He is the judge both of fact and of law and should therefore have a basic knowledge of the law of contract and of tort, together with a sufficient knowledge of the law of evidence to enable him to give just rulings upon the admissibility of the evidence presented.

TERMS OF THE APPOINTMENT

Whether chosen by agreement or appointed by an authority it is the arbitrator himself who chooses whether or not to accept the appointment. He may possibly require as a condition of his appointment that the parties agree to his scale of charges, but usually it is unwise for him to insist that both parties must agree. This is because refusal to agree may simply be a ploy in a reluctant party's strategy of delay.

Where the arbitrator has been chosen by agreement between the parties it is usual for the claimant to notify him of that agreement and to request that he accepts the appointment and proceeds with the reference. Where appointed by the President of the ICE or some other authority the arbitrator and the parties will be notified by the authority of the appointment.

The arbitrator should set out his proposals as to the terms of his appointment: usually as an hourly rate for time spent in his own office, and as a daily rate for attendance at meetings and hearings, plus travelling expenses, plus VAT. In addition he should define clearly the basis on which he requires remuneration for time allocated to meetings or hearings, but vacated (by the parties) before the event: usually as a percentage of his daily rate, which percentage increases as the period of notice of vacation is reduced. Such a provision is necessary in arbitration because many disputes are compromised before, and often immediately before, the date fixed for the hearing. A good example of a comprehensive agreement covering the terms of an arbitrator's appointment is that published by the Society of Construction Arbitrators: see appendix to SD/9.

The arbitrator should invite both parties to agree in writing to his proposed terms; and where both parties do so agree, no difficulty arises.

Where neither party agrees the arbitrator must either rely upon his power under section 18 of the 1950 Act to tax his own costs (subject to the safeguard against abuse of that power contained in section 19: see Chapter 9), or reject the appointment. In the former event he may well make his willingness to proceed conditional upon one or both of the parties providing security for his charges.

A problem that sometimes arises is that created by one party's agreement to the arbitrator's proposed charges, and the other party's rejection or failure to agree. In *K/S Norjarl* v. *Hyundai Heavy Industries (1990)* (TLR, 8 November 1990) it was held in the High Court that it was inappropriate for an arbitrator to conclude an agreement about fees with one party where the other party did not agree; because to do so might lay himself open to an imputation of bias. It is submitted therefore that where only one party agrees to his proposed terms of appointment he should proceed as though neither party had agreed.

Arbitrators sometimes require as a term of their appointment that they are empowered to take legal advice; the costs of any such advice forming part of the costs of the award. While such a term may be prudent as a means of covering unforeseen legal problems, it is submitted that in normal circumstances it is highly unlikely that the arbitrator would invoke such a power without seeking the parties' agreement, not only to his taking advice, but also to a definition of the question of law on which advice is to be sought, and as to the source of the advice. This applies in particular where both parties are represented by counsel.

The various ways in which an arbitrator may obtain guidance in determining questions of law are dealt with in Chapter 8.

SUPPLYING VACANCIES

It sometimes happens that after being appointed the arbitrator refuses to act (for example because he has become aware of some bar to his acting or for some other reason) or becomes incapable of acting, or dies. These situations are dealt with in Clause 66 of the ICE Conditions and more generally under section 10 of the 1950 Act, by which a party may serve on the other a notice to concur in the appointment of a replacement. If the other party fails to concur within seven days, the party who gave the notice may apply to the High Court to make an appointment.

A further difficulty that may arise, and one which was not foreseen in the

1950 Act, is that an appointing authority named in an arbitration agreement may neglect or refuse or be unable to make an appointment. This gap in the 1950 Act has been dealt with by way of an amendment to section 10, enacted as part of section 6 of the 1979 Act, which covers such failures and provides for appointment by notice to concur or by the High Court, as in other cases.

5 Preliminaries

SYNOPSIS

Before matters in dispute can be brought to a hearing the issues to be determined must be defined. Each party must be forewarned of the case it has to answer: relevant documents must be disclosed and where possible agreed. Arrangements as to representation and appointment of experts must be defined and notified: and arrangements for the hearing must be made.

ENGLISH LAW

It is a basic principle of English law that surprise tactics shall not be used, either in litigation or in arbitration. Neither party is allowed to gain an advantage by, for example, appointing leading or junior counsel to represent them at the hearing without giving notice of that intention, or by making some allegation that has not been disclosed to the other party and for which that other party has been unable to prepare. Neither party may without notice call an expert to give evidence, because without such notice the other party would be unable to appoint its expert – possibly to express a different opinion.

English law also requires that all documents relevant to the matters in dispute must be disclosed to the other party, whether or not they support the case of the party holding them. These two requirements of the law are dealt with during the preliminary or 'interlocutory' stages of an arbitration, and are termed 'pleadings' and 'discovery of documents'. But there are also other matters, such as special rules of procedure, and the timetable for the various stages, which may need to be defined: preferably at the beginning of the proceedings.

THE PRELIMINARY MEETING

Sometimes called a 'meeting for directions', the preliminary meeting is a meeting between both parties' representatives and the arbitrator, usually called by the arbitrator as soon as he has accepted his appointment, in order that procedures, arrangements and a timetable may be agreed and defined. Such definition is given in an 'Order for Directions' issued to both parties by the arbitrator after the meeting (see SD/16).

Alternatively an Order for Directions may sometimes be agreed in correspondence between the parties and sent as a draft to the arbitrator, who will then issue a formal Order by consent. Disadvantages of this procedure are that it provides no opportunity for the arbitrator to ascertain the nature and magnitude of the matters in dispute, of which he may at this stage have little if any knowledge; it provides no opportunity for the arbitrator to influence the parties into agreeing upon simplifying procedures and agreeing upon as short periods of time for the various stages as may be practicable, or for any discussion between the parties and the arbitrator; and the draft Order as prepared by the parties may not cover all of the matters that need to be defined. On the other hand it saves the time and cost of a meeting, and this may be an important consideration where the sum in dispute is small or where the parties and the arbitrator are in widely scattered locations.

In the first edition of this book a check list of matters that may need to be discussed at a preliminary meeting was included, in accordance with the then prevailing practice. It is however submitted that it is helpful to the parties to be forewarned of those matters, and accordingly that the check list should be replaced by an agenda, issued to both parties before the meeting. A typical agenda for a preliminary meeting is provided as SD/15. In addition there may be other matters to be considered: where, for example, the arbitrator's jurisdiction is challenged.

If he has not already done so the arbitrator should ensure at the preliminary meeting that his appointment is in order. He should check the contract and the arbitration clause therein; that the procedure for his appointment has been followed properly; that the subject matter of the dispute is within his own expertise but that he has no prior knowledge of the particular dispute with which he is to be concerned; and finally that he has no relationship or interest which might, or might be thought to, impair his impartiality. He may, if he has not already done so, seek the parties' agreement to the terms of his appointment.

The next step at the preliminary meeting is to define the parties' intentions as to representation and as to the appointment of experts, as these matters will affect the time needed for preparation of the pleadings, and the later stages of the arbitration.

Where both parties elect to appoint counsel the arbitrator will usually deem the proceedings to be 'fit for counsel' and will mark his Award accordingly. By so doing he will enable each party to include counsel's fees and expenses in its bill of costs at the conclusion of the reference. Where both parties agree not to appoint counsel there is no difficulty. But where one party insists upon appointing against the wishes of the other, it will be for the arbitrator to decide, probably later in the proceedings, as to whether or not the appointment was necessary, and to mark papers accordingly. In some cases the party not wishing to appoint counsel may feel obliged by the action of the other party to do so; in such a case the arbitrator should consider whether or not to require the party who insists upon counsel to bear both parties' counsels' costs in any event – that is, irrespective of the outcome of the claim.

At this stage it is appropriate to consider what, if any, special rules of procedure should apply to the arbitration. Certain rules may already have been agreed within the terms of the arbitration agreement: and where that is the case the arbitrator should ensure that both parties are aware of the existence and significance of such rules. The parties should consider whether or not the rules already adopted are appropriate to the known facts of the dispute; and where necessary should agree to modification or deletion of those rules. The adoption of new or additional rules may also be considered: see Chapter 2 and SD/15.

PLEADINGS

Pleadings have been defined as statements in writing served by each party alternately upon the other stating the facts relied upon to support its case and giving such details as the opposing party may need to know in order to prepare its case. Besides providing a means of forewarning each party of the case of its opponent, pleadings serve as a means of defining the questions of fact and of law that have to be determined by the arbitrator. Each document in the pleadings must state material facts only, in summary form, and not the evidence by which those facts are to be proved.

Documents in the pleadings comprise Points of Claim, Points of Defence, and Points of Reply (see SD/17, 20 and 21). The reply provides an opportunity for the claimant to deal with any allegations in the Points of Defence that are not mere rebuttals of the Points of Claim; the Points of Reply must not raise any new allegations.

Where there is a counterclaim – that is, a claim by the respondent against the claimant – it is pleaded with the Points of Defence, the words 'and Counterclaim' being added to the title, and details being given at the end of the defence. Thereafter the pleadings relating to the counterclaim follow one step behind those relating to the claim: the Defence to Counterclaim being pleaded with the Points of Reply, and where necessary there may also be a Reply to Defence to Counterclaim.

In order to facilitate reference to their content each of the pleadings should contain a series of numbered paragraphs each of which deals with a single subject. Points of Claim should commence with a brief introduction of the parties and should then identify the contract from which the dispute arises: its purpose and an outline of the respective responsibilities and rights of the parties. Next it is usual to identify and summarize the particular clauses of the contract that are relevant to the matters in dispute, and to state in what way those clauses – or where appropriate, terms that may be implied into the contract – have been breached. Finally the Points of Claim should state the amount of the claimant's entitlement to additional payment under the terms of the contract, and/or the loss suffered by the claimant as a result of breaches of the contract. Particulars of all such claims should be stated in order that the respondent may be aware of the amount of the monetary claim against him.

Damages may in this context be defined as the loss suffered as a result of the breach of contract; the legal principle governing their assessment is that the injured party should be put as nearly as possible in the same position as he would have been had there been no breach. There are two main categories: special damages, which are not presumed at law, but must be expressly pleaded and proved; and general damages, which follow naturally from the wrong pleaded, and need not therefore be pleaded.

Where the arbitration arises from rejection of a claim for additional payment, the damages claimed are likely to be wholly special damages, which must be pleaded and particularized in full. In such cases the breach complained of should be defined, and the special damages claimed should be set out in a schedule of particulars: where necessary there should be such a schedule for each breach pleaded.

The basic rule in drafting Points of Defence is that any allegation in the Points of Claim that is not denied or otherwise dealt with is deemed to be admitted. There are three ways in which the respondent may deal with an allegation in the Points of Claim: he may admit, deny, or 'not admit' it.

In most cases there are several statements in the claim that are not in dispute: for example the existence of the contract, its purpose, and the content of clauses in it. In such cases the Points of Defence should admit those statements: but it may at the same time be desirable to draw attention to other documents, or clauses in the contract, which have not been quoted in the Points of Claim because they do not favour the claimant's case.

Secondly, there may be allegations in the Points of Claim which the respondent believes to be untrue, and these must be denied. Thirdly, there may be allegations relating to matters of which the respondent has insufficient knowledge either to admit or to deny. Where such allegations are important to the claimant's case the respondent may state in the defence that they are 'not admitted' – thereby requiring the claimant to prove them.

Besides dealing with each of the allegations in the Points of Claim the Points of Defence may include fresh allegations; for example, in a case in which the claimant alleges that his work was delayed through some fault of the employer or of his engineer, an allegation that the delay was caused by some inadequacy in the contractor's organization. Finally, it is usual in most cases to include a 'blanket denial' of all allegations not specifically admitted (see paragraph 8 of SD/20) to cover the possibility that some of the items in the Points of Claim may not have been dealt with adequately.

In some cases a party may consider that the pleadings are not sufficiently detailed to give that party proper forewarning of the case to be answered. In such a situation the party should serve a Request for Further and Better Particulars (see SD/8).

AMENDMENT OF PLEADINGS

During the course of the pleadings, during later stages of the interlocutories, or even during the hearing, a need may arise to amend pleadings, for example when the Points of Defence disclose some fact not previously known to the claimant, or where fresh evidence comes to light during discovery or during the hearing. In such cases the party wishing to amend its pleadings should apply to the arbitrator for his consent, giving reasons for the application, and of course sending a copy to the other party. Generally, such a request, if soundly based, will be granted, with the proviso that the applicant shall bear any costs that may result from the amendment, in any event; that is, irrespective of the outcome of the arbitration.

There are however exceptions to this general rule. Where for example the need to make the amendment arises from some fault on the part of the other party, the party making the application should bring the fault to the notice of the arbitrator, and request that he exercise his discretion as to costs (see Chapter 9). If there is a dispute between the parties as to who should bear the costs of the amendment, the arbitrator may possibly reserve his decision on the point, or may call a meeting at which he will hear both parties' contentions before coming to a decision.

The amount of costs that may result from an amendment of the pleadings may vary widely according to the extent of the amendment and the stage at which it is allowed. In most cases, where the application to amend is made during the pleadings or during discovery, the amendment will result in a need for subsequent pleadings also to be amended; the costs of such amendments will fall upon the applicant for the original amendment, subject to the exceptions referred to above. An application to amend made during discovery is likely, in addition, to involve delay; while an application to amend pleadings made during the hearing will, if granted, usually require adjournment of the hearing and will therefore involve the costs of time allocated to it by the parties and their representatives and witnesses and by the arbitrator. Such a situation may arise where a party alleges, during the hearing, that evidence being presented by his opponent relates to matters that were not disclosed in the pleadings. If the arbitrator upholds the allegation, then he may ask the party giving evidence if he wishes to apply to amend his pleadings; otherwise he must require the party to abandon the evidence in question. In considering which course to adopt the amount of costs likely to be incurred if the amendment proceeds, may be an important factor affecting the applicant's decision.

THE SCOTT SCHEDULE

Where a claim comprises a large number of items each of which has a separate basis in the contract, it is often convenient for the arbitrator to have the pleadings summarized in the form of a Scott Schedule, sometimes called an Official Referee's Schedule. Such a schedule usually requires preparation by both parties, and does not have any fixed format, other than the basic principle that each item is taken separately, and includes the contentions of both parties in relation to that item (see SD/22).

DISCOVERY OF DOCUMENTS

After the close of pleadings, each party must prepare lists of documents (see SD/23) that are or have been in the party's 'possession or power' relating to the matters in dispute, and must serve a copy of the lists upon the opposing party. All documents listed, other than those that are 'privileged', must be made available for inspection by the other party, and if required, copies must be made at the expense of the party requiring the copies. On completion of discovery and inspection the parties should compile an agreed list of documents, being an amalgam of both parties' lists, and should ensure that each document is serially numbered.

The requirement to disclose all relevant documents is observed strictly, and applies not only to those documents that support the case of the party holding them, but also to documents that detract from that case. The principle is that the arbitrator requires to discover the truth; any document that helps to throw light on that truth, however embarrassing it may be, is of value to the arbitrator. Thus for example a contractor who is seeking enhanced payment in respect of an item of varied work must expect to have to disclose pricing notes showing the make-up of the original rate. Again, correspondence between the employer and the main contractor may well be relevant to a dispute between main and subcontractor, especially where inadequacy in the subcontractor's work is alleged.

Usually the relevant documents in a dispute arising from a construction contract comprise all of the contract documents and drawings, all correspondence between the parties and between the contractor and the engineer, including drawings, instructions, variation orders and certificates issued during the course of the works; site diaries; interim valuations submitted by the contractor during the course of the works; the final account; the contractor's office records of wages, plant, oncosts, daywork sheets, pricing notes, internal memos and correspondence between the site agent and his head office. Additionally there may also be relevant correspondence between the contractor and third parties such as sub-contractors, suppliers and technical advisers.

Where a party suspects that not all relevant documents have been discovered by its opponent the party may serve a notice requiring an affidavit verifying the list of documents, or requiring a statement as to what has become of a document known to have existed.

Usually a dispute between the parties arising from an allegation of incomplete discovery is resolved between them, but if necessary a party may apply to the arbitrator for an order requiring the opposing party to discover the documents in question. Failure to comply with such an order would provide grounds for the aggrieved party to apply to the High Court under section 12(6) of the 1950 Act for an order requiring discovery.

PRIVILEGE

Certain types of document need not be produced on discovery, although they must be included in the list of documents, in a separate schedule marked to indicate that privilege is claimed. Privilege relates to correspondence between clients and their legal advisers; more generally to any letter or document prepared with a view to litigation or arbitration, and to offers to settle the dispute, where marked 'without prejudice'. It is sometimes claimed that certain documents should be privileged because they contain confidential information. Such claims can rarely be upheld, because arbitration proceedings are private, and confidentiality, which might otherwise be a factor, is of minor importance. In *Mitchell Construction Kinnear Moodie Group* v. *East Anglia Regional Hospital Board (1971)* (Current Law Yearbook 375) personal files relating to the contractor's employees were ordered to be disclosed, it being held that the sole issue was one of relevance. Similar principles would be applied to the discovery of other private documents, internal memoranda, pricing and estimating notes, where they can be shown to be relevant to the matters in issue.

Agreed bundles of documents

A difficulty often arises, both in arbitration and in litigation, from the enormous volume of papers generated by a construction project.

If all documents that have been issued in connexion with the contract are deemed to be relevant in the sense that they relate to the contract, then a set of 'relevant' documents may well comprise, in a major dispute, perhaps twenty or thirty lever arch files (or 'bundles', as they are usually termed) each containing several hundreds of papers. For the purpose of the hearing four sets of bundles are required: one for the arbitrator, one for the witness giving evidence, and one for each counsel. Of the papers so prepared it is likely that in many cases only a very small proportion will be referred to during the hearing.

In order to limit the considerable cost that may be involved in listing, paginating and copying numerous documents many arbitrators now direct that only those papers that are relevant to the identified issues should be included in agreed bundles. This procedure may reduce the volume of paper to be prepared, perhaps to 1% of the total number of documents; but it has the disadvantage of requiring time to be spent, by persons who have knowledge of the issues and are therefore relatively senior, in identifying the relevant papers. The alternative, of including all of the documents, requires the services of only the most junior clerk in preparing the bundles comprising all documents in the case.

A compromise often adopted is to prepare in quadruplicate 'core bundles' of documents identified as being important, for the participants in the arbitration, and to have all other papers available, but not copied, for reference should the need arise.

Whichever procedure is adopted the bundles are denoted by letters of the alphabet, and the papers within them are numbered serially.

The use of electronic storage and retrieval systems as a means of obviating the cost and inconvenience of extensive copying is in its infancy; but it is to be expected that such systems will come into general use in the foreseeable future. A necessary prerequisite of such general adoption is agreement to a standard system; and it is perhaps that requirement that presents the main obstacle in a rapidly developing field of technology.

Agreement of figures as figures

An Order for Directions usually includes an order that where possible figures shall be agreed as figures.

The purpose of such an order is to reduce the need for detailed evidence during a hearing, where such evidence can be, and usually is, capable of being agreed between the parties' quantity surveyors or other technical representatives. The figures referred to include, but are not confined to, monetary matters: they may also include records of numbers of workmen or items of plant employed on the site; numbers of days on which weather was inclement; the time taken to perform sections of work, or any other figures that might otherwise be a subject of contention. A private meeting between the parties' quantity surveyors, besides being more conducive to agreement, is likely to be far less expensive than hearing time at which all of the parties' representatives and lawyers, and the arbitrator, are present.

TRANSCRIPT OF THE HEARING

If the arbitrator thinks it necessary he may seek the parties' agreement to there being made a transcript of the hearing, or of part of it (for example, the closing addresses by both counsel), but he may not order such a transcript without the consent of both parties. The cost of attendance of a shorthand writer, and of transcribing, may be a major item in the total costs of the reference, and for this reason many arbitrators are content to rely upon their own notes, possibly supplemented by a tape recording of proceedings at the hearing. In such cases the arbitrator will himself make the recording, taking a careful note of tape numbers and recorder meter readings at important events, such as the start of each witness's evidence. This will enable him to refer without undue difficulty to any passage in the evidence, so that he can ascertain the actual words used by the witness where contention arises.

ARRANGEMENTS FOR THE HEARING

Unless the arbitration agreement determines otherwise, the arbitrator has absolute discretion as to the date, time and place of the hearing. He should however exercise that discretion reasonably, bearing in mind that one of the reasons for the choice of arbitration is that it seeks to serve the convenience of the parties.

Where a large number of witnesses are required, and especially where solicitors and counsel are appointed, choosing a date or dates suitable to everyone concerned may present difficulty. For this reason it is usually desirable to fix at least a provisional date for the start of the hearing either at the preliminary meeting or soon afterwards, so that dates may be reserved by everyone concerned. By raising this point at the preliminary meeting the arbitrator can hear the parties' wishes and can comply with them as far as is reasonable and possible.

The location of the hearing may have been determined in the arbitration agreement; but whether or not this is so the parties are free to agree upon a location best suited to the convenience of all of those concerned. In major cases this often means London, because many counsel and major firms of solicitors are based there; other locations may be considered upon their merits. The arbitrator will be wise to ensure that the chosen location is one in which the procedural law of the arbitration applies: meaning, for example, a location in either England or Wales if the arbitration is to be conducted in accordance with English law.

The venue of the hearing should be a courtroom, hall or office sufficiently large to accommodate all of those involved in reasonable comfort, and having regard to the probable need to refer to and perhaps to display large numbers of documents and drawings. The usual layout of the courtroom provides two long and parallel tables, one for each party, and a linking table across one end at which the arbitrator sits. Witnesses, when called, sit facing the arbitrator between the two parallel tables, so that their evidence and demeanour may be heard and seen by both parties and by the arbitrator. In smaller or less formal hearings the parties may sit on either side of a single large table, with the arbitrator at the head, and with witnesses when called either at the foot of the table or in whatever position enables them to be best heard and seen.

In some cases, and especially in major arbitrations and in those in which counsel are appointed, the parties require retiring rooms where they may confer during the lunch and other intervals.

Suitable courtrooms for arbitrations are available on hire at the Institution of Civil Engineers, the Chartered Institute of Arbitrators, and at other locations.

Accommodation for the hearing is sometimes booked by the arbitrator and sometimes by one of the parties: probably the claimant. In some cases a party may offer the use of its premises for the hearing: such an offer should not be accepted by the arbitrator without the agreement of the other party.

CONDUCT OF INTERLOCUTORY STAGES

The arbitrator should at all times exercise the greatest care in ensuring that he is, and that he is seen to be, impartial in his dealings with both parties. He should not have any communication with a party without the knowledge of the other; this in practice means that he either writes letters addressed to both parties, or when writing to one party sends a copy to the other. He should ensure that a copy of any letter he receives from one party is sent to the other, either by the writer or by himself, and he should avoid having any communication by telephone. Although in the past arbitrators sometimes relaxed that rule to allow for discussion by telephone of dates for meetings, the advent of the fax machine has obviated the need for such a relaxation which was highly undesirable. Arbitrators may now require that any urgent communication be sent by fax, with a copy, also by fax, to the opposing party.

For similar reasons he should avoid having any meeting with a party except in the presence of the other: and he should not accept hospitality from either party, even in the presence of the other.

'LIBERTY TO APPLY'

Most Orders for Directions contain an order that there shall be liberty to apply. This phrase originates from practice in the courts, in which it means that the parties may come to the court again without taking out a fresh summons. Hence, in arbitration it may not be strictly necessary, but it serves to indicate that a party may where necessary apply to the arbitrator for an extension of time, for permission to amend pleadings, for a meeting or hearing, or for an order requiring the other party to take any action that may be needed during the interlocutory stages.

PRELIMINARY QUESTIONS OF LAW

It sometimes becomes clear during the interlocutory proceedings that the success of a claim, or of a substantial part of it, depends upon a question of law which, if determined as a preliminary matter, might save unnecessary delay and costs. Where for example it is pleaded that the circumstances of a claim bar it from consideration because of the operation of the Limitation Acts, such a defence, if successful, would obviate the need to proceed further.

The arbitrator has jurisdiction to determine questions of law, and he may do so either from his own knowledge or, where appropriate, he may take legal advice on the issues involved – provided that by so doing he does not delegate his authority to another person. Either of these courses will however leave the possibility that the question may be determined incorrectly.

Provision is made in the 1979 Act for such questions to be determined by the High Court, subject to compliance with certain requirements. Under section 2 of the Act, an application to the court must be made by one of the parties, either with the consent of the arbitrator or with the consent of the other party (or parties, where there are more than two). The arbitrator is not himself empowered to make such an application, although in appropriate cases he could suggest this course of action to the parties and indicate that he would give his consent to an application by one of them if this consent should be needed.

In considering an application under this section the High Court requires to be satisfied that determination of the question of law might produce a substantial saving in costs, and secondly that the question of law is one in respect of which leave to appeal would be likely to be given under section 1(3)(b) of the 1979 Act. The criteria applied by the High Court in considering whether or not to give leave to appeal under that section are dealt with in Chapter 10.

Applications under this section are dealt with in the Commercial Court of the Queen's Bench Division, and it is to be expected that they will normally be dealt with as matters of urgency.

SMALL CLAIMS

It is reasonable in this context to define small claims as those in which a risk exists that costs may be substantial in relation to the sum in dispute. The objective of procedures designed to deal with such claims is to ensure that costs are not allowed to become disproportionate to the claims; hence the definition is itself flexible. However as a very rough guide it is suggested that any claim amounting to less than six figures is potentially within this definition.

In many such cases the parties will not have appointed legal or technical representatives at the time of commencing the arbitration, and they may have little knowledge of their rights and duties during the reference; still less of their rights to agree with the other party upon simplifying procedures. The arbitrator should therefore make it clear to both parties that they have the right to be represented – provided that the arbitration agreement does not specify otherwise – upon giving notice of this intention, or to conduct their own cases. He should also make it clear that the costs of such representation may be substantial, and that the question who should ultimately bear the costs lies within his discretion.

On the assumption that one or both parties may opt not to be represented, and that that party or parties may not be familiar with arbitration procedure, the arbitrator may think it desirable to outline such procedure: an example of a letter setting out the basic information is given in SD/10.

Where the sum in dispute is so small as not even to warrant a hearing the arbitrator may, if he is satisfied that justice can be achieved without one, suggest that the parties agree to a 'documents only' procedure, combined if necessary with an inspection of real evidence. An example of a letter suggesting such a procedure is given in SD/11.

The adoption of any such procedure is however a matter requiring the agreement of *both* parties, since in making such an agreement each party is waiving its basic right to an oral hearing and to whatever form of representation it may choose. Where a party unreasonably rejects such a suggestion by the arbitrator, he may well take account of that rejection when he comes to exercise his discretion as to the award of costs.

6 Evidence

SYNOPSIS

Evidence is defined in *Osborn's Concise Law Dictionary* as being '. . . all the legal means, exclusive of mere argument, which tend to prove or disprove any matter of fact, the truth of which is submitted to judicial investigation'. The law of evidence is highly complex; while the rules deriving from the law need not be applied rigidly in arbitration, and may be varied by agreement between the parties, the principles upon which the rules are based should not be abandoned without good cause. In this chapter the kinds of evidence and rules relating to its presentation are outlined.

STATUTES

The principal statutes governing the law of evidence are the Evidence Act 1938, the Civil Evidence Act 1968, and the Civil Evidence Act 1972. The 1968 Act relates mainly to rules governing the admissibility of hearsay evidence, while the 1972 Act deals with expert evidence.

KINDS OF EVIDENCE

The main types of evidence adduced in construction arbitrations are documentary, oral and real evidence. There are however several other ways in which different kinds of evidence may be distinguished: for example, it may be evidence of fact or of opinion; it may be direct or circumstantial; it may be primary or secondary. Hence any item of evidence may be categorized in several different ways, depending upon the manner in which it is adduced, and upon its content and quality.

Documentary evidence

Under this heading is included all evidence in writing or on paper: letters, documents, drawings, photographs, computer print-outs and the like. All documents that are relevant to the matters in issue must be listed on discovery (see p. 36), including those for which privilege is claimed, and those which are not privileged must be made available for inspection and for copying.

In many construction arbitrations documentary evidence is by far the most important factor in determining the matters in question, especially where the parties have maintained comprehensive records and where a long interval elapses between events giving rise to the dispute and the arbitration.

In general the documents produced in evidence must be originals: the letter sent from one party to another, not the file copy retained by the sender. It should be noted however that the manner in which the document was originally prepared – whether as the original typed letter, as a carbon or photocopy or by any other means – is immaterial; the letter that was signed and sent is the original, even though it may have been prepared by photocopying.

Where the original document has been lost or destroyed a copy may be admissible.

It is a duty of the arbitrator to ensure that, where required under the Stamp Act 1891, documents taken in evidence must be adequately stamped, and he should not accept such documents in evidence until any deficiency in stamping has been corrected.

Oral evidence

The arbitrator is empowered, under section 12(3) of the 1950 Act, to administer oaths or to take affirmations of the parties and of their witnesses, and usually that power is exercised. A witness who knowingly makes a false statement under oath is guilty of perjury, which is punishable by fine or by imprisonment, or both.

Where necessary a party to a reference may compel a witness to attend the hearing, by means of a writ of *supoena ad testificandum*; and a Master of the High Court is empowered to issue such a writ to compel the attendance of any witness living in the United Kingdom.

Where evidence is required from a person who cannot readily be brought to the hearing a party may apply to bring evidence on affidavit, that is, a sworn statement; where necessary an order in respect of such evidence may be obtained from the High Court. Such evidence is however of less value than that given by witnesses present at the hearing – notwithstanding that they too may have provided evidence on affidavit – because they are available to submit to cross-examination.

Real evidence

Material objects presented for examination are termed real evidence. They may be samples brought to the hearing, such as bricks, pieces of concrete, steel sections and so forth, or they may be immovable objects such as bridges or dams, which have to be inspected on site. Where necessary the arbitrator may either before, during or after the hearing arrange to inspect real evidence on site, in the presence of both parties' representatives, who should be permitted to draw matters to his attention but not to make any representation to him.

Evidence of fact

Unless a witness is qualified as an expert (see below) his evidence must be of factual matters only, and of matters that are within his personal knowledge. He is not permitted to express opinions.

In many cases however the facts upon which opinions are based are of more evidential value than the opinions themselves: for example a statement of the dates on which drawings were issued in relation to the programme for construction of the works shown on them is likely to carry more weight than an opinion that the drawings were issued late. Similarly a statement that a brick wall is 30 mm out of plumb and contains courses varying from 72 to 77 mm in height is of more value than a subjective opinion that the workmanship of the brickwork is of poor quality.

The Civil Evidence Act 1972 does however relax the rules relating to evidence of opinion, as outlined in the following paragraphs.

Expert evidence

A witness may be called to give evidence of opinion on matters in which he is suitably qualified. Such a witness is termed an expert and his appointment and evidence is subject to special rules, principally those contained in the Civil Evidence Act 1972. The role of an expert is to assist the arbitrator in coming to a correct decision on the points for which he is called to give evidence: he is not in any sense an advocate for the party calling him. For this reason an expert should not accept an appointment as such until he is sure that the evidence he is able to give will support the case of the party proposing to call him. Having satisfied himself on this point, and having accepted the appointment, he must still direct his evidence to elucidation of the truth rather than the mere presentation of his party's case. This does not imply that he should make the opposing party's case for him: but he should be aware of any weaknesses in his evidence and should give it with honesty and sincerity.

An expert's qualifications as such need not necessarily be formal, such as degrees of universities or membership of learned bodies. In some cases experience may be the prime requirement: for example where an issue arises from a claim under Clause 12 of the ICE Conditions as to whether or not conditions or obstructions encountered by the contractor were such as might reasonably have been foreseen by an experienced contractor, the appropriate expert would be an experienced contractor.

The rules as to who may give evidence of opinion have been relaxed under the Civil Evidence Act 1972 and now permit a person called as a witness of fact to give a statement of opinion on any relevant matter on which he is qualified. This situation often arises in construction disputes in that a witness of fact, such as the contractor's agent or the resident engineer, is qualified to give evidence of opinion on technical questions. Furthermore, a witness of fact may also give a statement of opinion on matters for which he is not qualified, where that statement conveys facts perceived by him personally.

Hearsay evidence

One of the rules of evidence is that statements made by another person to a witness are not admissible unless the other person is the opposing party or his agent, or unless the statement was made in the presence of the opposing party or his agent. In general, the maker of the original statement should be called to give the evidence; and even where the rules allow hearsay evidence to be admitted it is of less weight than that of the originator of the statement.

This basic rule has however been relaxed under the Civil Evidence Act 1968 which allows hearsay evidence to be admitted by leave of the court, subject to certain rules. These permit, for example, hearsay evidence of what a witness said on some previous occasion to be admitted after that witness has given his evidence-in-chief, thereby allowing inconsistencies in the witness's evidence to be brought to light.

ADMISSIBILITY OF EVIDENCE

The primary requirement of evidence in order that it may be admitted is that it must be relevant to the points at issue. It is for the arbitrator to decide any question as to admissibility, but generally, where the parties are legally represented, he need not reject any evidence unless an objection is raised by the other party. Where a party is not so represented however, he may think it necessary to draw attention to any defect in the evidence being presented by the other and to invite an objection.

The arbitrator should not refuse to hear any admissible evidence: but should it appear to him that a point has already been adequately covered and is being laboured unnecessarily, he may suggest to the party bringing the evidence that he has already heard enough evidence on that point.

PROOFS OF EVIDENCE

A witness of fact is not permitted to read from a prepared statement, or even to refer to such a statement, while giving evidence. He is, however, permitted to refer to notes taken contemporaneously with the events referred to – for example his diary.

In order to assist the solicitor or counsel conducting his examination-in-chief it is usual for the evidence of each witness to be called to be set down as a proof of evidence, to which counsel will refer while examining the witness. By skilful questioning he will bring out all of the points that need to be covered by the witness; but he must not lead his witness (see p. 52).

In recent years, however, the practice of exchanging between parties their proofs of evidence of witnesses of fact has been increasing in popularity. The advantage of this procedure is that it saves the time that would otherwise be taken in laborious questioning of the witnesses by their own counsel, and recording of questions and answers in manuscript by the arbitrator. While it may be argued that the practice provides an opportunity for an unscrupulous witness to prepare and check at his leisure an untruthful account of the events witnessed it is unlikely that such an account would survive under skilful cross-examination. Counsel conducting such cross-examination would have the benefit of his own party's version of the events in question, and the availability of the untruthful evidence in writing before the start of the hearing should greatly assist him in preparing his questions.

An expert witness is permitted to refer to his report or other statement prepared by him after the event, while giving his evidence. Nevertheless it is usual for counsel to prepare a proof of expert evidence in a similar way to that of a witness of fact, so that counsel may ensure that important points are covered and emphasized as necessary.

An example of a proof of evidence is given in SD/25.

CALLING WITNESSES

The arbitrator is not empowered to call witnesses; only the parties may do so. Should it appear to the arbitrator that some person ought to be able to give evidence on the matters in issue, or that some point may not be covered, he can bring the omission to the notice of the parties by an enquiry whether, for example, the engineer under the contract is to be called, or whether he will be hearing evidence on the point that needs to be covered. If he finds that the point needs to be dealt with by a person who has already given evidence he can ask the party in question to recall that witness for further questioning.

BURDEN AND STANDARD OF PROOF

The burden of proving an assertion made by a party lies with that party. Having provided such proof, the burden shifts to the other party who will have either to disprove the assertion, or show that it is not a material point in dealing with the main issues, in order to return the burden again to the party making the assertion.

In arbitration, as in civil actions in court, the standard of proof required is that of *balance of probability*. If it appears to the arbitrator more probable than not that events were as stated by a party he will find in favour of that party. This standard may be contrasted with that required in the criminal courts, in which the prosecution must prove guilt *beyond reasonable doubt* in order to obtain a conviction.

7 The hearing

SYNOPSIS

Rules governing attendance at the hearing; privacy; courtesy; challenges to jurisdiction; procedure and sequence.

ATTENDANCE

Notwithstanding that the arbitrator may have agreed the venue, date and time of the hearing with the parties, he should issue a formal notice to them giving this information and requiring them to acknowledge receipt of it. By this means he is able to ensure that there is no valid excuse for a party's absence: although if a party does not appear he should generally wait a reasonable period of time to allow for possible delays in travelling, before adjourning the hearing. In that event he should fix a new date and time for the hearing, notifying both parties as before, and adding the warning that in the event that either party fails to attend he will proceed with the hearing *ex parte* (see p. 23).

In arbitration the hearing is private; attendance is limited to the parties and their representatives and witnesses, the arbitrator and, where agreed by the parties, a legal adviser to the arbitrator and possibly a shorthand writer. Others may attend only with the consent of both parties and of the arbitrator, and usually such consent is given where a student wishes to attend, but not where the press or members of the public with no special interest apply to attend.

Where a question arises as to the credibility of one or more witnesses they may be excluded from the hearing until required to give their evidence.

COURTESY

In a formal hearing the arbitrator may arrange for a member of his staff to ensure that both parties are present in the courtroom before he enters. In such a case those present rise when he enters, and sit only when invited to do so.

The arbitrator should be addressed as 'Sir' or, when referred to in the third person as 'The Arbitrator' – not as 'Mr Smith'.

Except while administering the oath, the arbitrator usually remains seated during the hearing. Counsel, solicitors and witnesses – again except while taking the oath – may also remain seated, although some counsel prefer to stand.

CHALLENGES TO JURISDICTION

It may happen at the commencement of a hearing that a party challenges the arbitrator's jurisdiction as arbitrator. In such an event the arbitrator should hear the grounds upon which the challenge is based, and should ascertain why it is raised at this stage and not during the preliminaries. He should also provide an opportunity for the opposing party to make any submission that may be relevant to the matter. The arbitrator will then have to decide whether to proceed with the hearing, to adjourn it, or to abandon it altogether. Should he decide to proceed he will note the substance of the objection raised and advise both parties of his ruling, thereby leaving it open to the challenger to take further action should he so decide.

If the arbitrator thinks it prudent to adjourn the hearing while the challenge is investigated, he should warn the parties that taking this course will inevitably result in additional costs because of the time wasted by himself and by the parties and their representatives and witnesses; and that he will in due course award such additional costs having regard to the outcome of the challenge. Such a warning may, in cases where the challenge is of doubtful validity, result in its being abandoned.

REPRESENTATION

The parties are free to choose the form of their representation, if any; they may appoint a layman, a technical advocate, a solicitor or counsel; but as explained in Chapter 5 they must make their intentions known to the other party.

If a party appears at the hearing with counsel without having given notice of that intention the arbitrator should invite the other party to apply for an adjournment to enable it to arrange similar representation if it so wishes. In such an event the additional costs resulting from the adjournment would be awarded against the party who failed to give notice.

PROCEDURE

The arbitrator opens the hearing by reciting in outline the events that have led to it: the identity of the parties and of the contract between them, the arbitration agreement therein, the occurrence of a dispute, the manner of appointment of the arbitrator and his acceptance of that appointment. He then declares the hearing open.

Usually* the claimant's counsel or representative, or the claimant himself where not represented, opens his case by introducing himself and others of his party, and he may also introduce counsel for the defence where so represented. Where counsel has prepared and submitted a written opening he will merely refer to that document and ask the arbitrator if he has had an opportunity to read it, and if there are any matters that require explanation. He may also take the opportunity to emphasize any parts of the statement he considers to be especially important: but the arbitrator should discourage any prolonged oral address which defeats the intention of the written submission.

Where this practice has not been adopted counsel will have to outline the claimant's case, drawing attention in particular to the matters in dispute between the parties, and to the evidence he will be calling to deal with those matters.

Where, as is usual, evidence is to be given under oath or affirmation, the arbitrator will have prepared himself by having available a Bible, an Old Testament, and a card on which is typed the form of oath and of affirmation, as follows:

> 'I swear by Almighty God that the evidence I shall give touching the matters in dispute in this reference shall be the truth, the whole truth, and nothing but the truth.'
>
> 'I solemnly, sincerely and truly affirm and declare that I will true answers make to all such questions as shall be asked of me touching the matters in difference in this reference.'

Witnesses who are Christians should be required to stand and to swear upon the Bible, holding it in their right hand while taking the oath. Those of the Jewish faith should repeat the same words, holding the Old

* In some cases, where the burden of proof lies initially on the respondent, the respondent opens. This happens, for example, where the claim is admitted, but it is alleged that it is nullified by the counterclaim, which the claimant denies. The respondent is said in such cases to 'confess and avoid' the claim.

Testament. Others should be required to affirm by standing and repeating the words of the affirmation.

The sequence of giving evidence is firstly examination-in-chief by counsel for the witness's own party, cross-examination by the opposing counsel, and finally re-examination by the witness's own party. A witness of fact must not, as explained in Chapter 6, refer to notes while giving evidence, other than those made contemporaneously with the events he is describing. Should he refer to any written record the arbitrator should ask to see it and should allow the other party an opportunity to do so, to check that it is a genuine diary or other note made at the time.

However, it has in recent years became standard practice in most arbitrations for the parties to exchange written and signed proofs of evidence of witnesses of fact (see Chapter 6). Where this practice has been followed, the need for examination-in-chief is eliminated. Counsel will, having identified each witness, refer him to his statement and ask if it is true; and if the witness wishes to add anything to it. In some cases the witness may wish to comment on any matter that has come to light in earlier evidence, but he should not have to expand or elaborate his proof if it has been properly prepared.

Where the witness finds it necessary to add to his written statement matters which ought to have been included in it the opposing counsel may draw to the arbitrator's attention a failure to comply adequately with his direction regarding the exchange of proofs of evidence, and may request that that failure be taken into account in the arbitrator's award of costs.

A witness must not be 'led' during examination-in-chief or during re-examination: that is, he may not be asked questions which suggest their answer. He must not be asked 'was the concrete sloppy?' but may be asked 'what was the consistency of the concrete?'. In practice, however, leading questions are usually permitted during examination-in-chief where they relate to matters that are not in contention: the identity of the witness, his appointment in his work and his presence on the site. This is simply to save time: it is quicker for counsel to obtain the answers he needs by asking: 'you are Joe Bloggs, you live at 25 High Street, Blanktown, and you are employed as a site engineer by Bill Smith?' than to ask each of the questions that would otherwise be needed. The same principle is sometimes extended to other questions; but as soon as contentious matters are dealt with, the proper form of questioning must be used.

Usually the arbitrator will not object to the form of questioning where both parties are represented by lawyers, because they should be aware of the rules, and are free to object where necessary. Where the parties are not so represented, however, the arbitrator may require them to comply with the rules.

Leading questions may be asked during cross-examination: indeed they are often necessary, because one of the duties of cross-examining counsel is to put his version of events to witnesses called by his opponents. For example a witness for a contractor may be asked during cross-examination: 'I put it to you that the delays were not caused by late issue of the drawings, but by shortage of labour on the site?' It must be borne in mind that each witness will, or should, be called once only; that during that appearance he must be given the opportunity to comment upon the opposing party's version of the facts.

Questions during re-examination must be confined to matters that have been raised during cross-examination, usually with the object of correcting any misleading impressions that may have been given by the witness during that cross-examination. In the event that such re-examination strays – perhaps unintentionally – into matters that have not previously been raised, the opposing party's counsel would probably ask for, and be allowed, an opportunity to cross-examine on those new matters.

The arbitrator may ask questions of witnesses and indeed he should do so where he finds that some matter has not been dealt with adequately or at all, or where he finds a witness's evidence to be inconsistent with his own knowledge and/or experience. In some cases the arbitrator may have to address his questions to counsel: where for example a witness does not give evidence dealing with matters that the arbitrator expected him to cover, he may ask counsel, 'will you be calling evidence as to the cause of the delay?'. Such questions are helpful to counsel in that they may indicate lacunae in the evidence; and they provide an indication of the arbitrator's thought process. Having been appointed as a person with expert knowledge of construction the arbitrator has a duty to use that expert knowledge, not only in weighing the evidence presented by the parties, but also in drawing attention to any need there may be for additional evidence to elucidate relevant facts. Where, as sometimes happens, the arbitrator by using his expert knowledge formulates an opinion in his own mind, he should be careful to expose it to the parties in order that they may comment upon or challenge that opinion. In *Fisher and Another* v. *P. G. Wellfair (1981)* (19 BLR 52) (see p. 57) the arbitrator's failure to do so was held to constitute misconduct.

Wherever possible the arbitrator should, however, ask his questions at the conclusion of re-examination, in order not to interrupt the flow of evidence during counsels' examination. Where the arbitrator's questions bring to light some new evidence he should if necessary permit further questioning by either or both counsel.

Reference may be made, while taking oral evidence, to documents produced in evidence. Wherever possible such documents should have been agreed between the parties, copied and formed into bundles, each document being identified by a bundle letter and a number. In this way both parties and the arbitrator may refer without difficulty to such documents, or to items of real evidence. Similarly where figures have been agreed 'as figures' the agreement should have been put into writing and the letter incorporated in one of the agreed bundles.

Except where both parties have agreed to the making of a transcript of the hearing the arbitrator must rely upon his own notes, possibly supplemented by a tape recording. He will accordingly require evidence to be given sufficiently slowly to enable him to keep an adequate record. Where counsel are involved they wait until the arbitrator has finished writing his note of an answer before proceeding to the next question.

After each of the claimant's witnesses has given evidence-in-chief, has been cross-examined and re-examined, and has answered any questions the arbitrator may ask, that witness is allowed to stand down. The next witness is called and questioned in the same sequence, after having taken the oath, and this procedure is repeated until all of the claimant's witnesses have been called.

Counsel or other representative for the respondent is then invited to open his case, and may do so either by outlining the substance of the defence, or by calling his first witness. In many hearings the case for the defence will already have been made clear during cross-examination of the claimant's witnesses, so that an opening address may not be necessary.

Each of the defence witnesses is examined-in-chief, cross-examined and re-examined, and may be asked questions by the arbitrator, repeating the sequence of evidence for the claimant.

While giving evidence a witness must not have any private communication with his party's counsel, solicitors or any other witness. Where possible opportunities for such communication are eliminated by arranging for the whole of the witness's evidence to be given during a single sitting of the tribunal; and where necessary sessions are often extended so that the evidence of a witness may be completed. In some cases, however, witnesses may have to give evidence for a period extending over several days; and where that happens the arbitrator should warn the witness whenever a break occurs that he must not communicate with his colleagues, or with any other person in respect of the matters relevant to the arbitration, during the recess.

Expert witnesses may be called at any appropriate stage of his party's case. It is to be expected that experts of like disciplines will have prepared an agreed report covering part, if not all, of the technical issues. If the experts are fully in agreement, there is no need for them to be called; they merely submit, jointly, the agreed report. Where their agreement does not cover all of the matters referred to them each expert will have prepared his individual report on those remaining matters, will be called, examined-in-chief where necessary, and may be cross-examined and re-examined as in the case of witnesses of fact.

When all of the defence witnesses have given their evidence, the defence counsel or representative makes his closing speech, in which he emphasizes the merits of his client's case and detracts from that of his opponent. Finally the claimant's counsel or representative makes his closing speech, again urging acceptance of his client's case. In a long and complex case both counsels may find difficulty in making their closing addresses immediately after the conclusion of the evidence. For this reason it may be desirable to allow an interval between these two stages, or, where appropriate, to adjourn the hearing until the following day. Alternatively it may be desirable to modify the procedure in order to provide for closing addresses to be submitted in writing (see p. 11).

It may sometimes happen that the respondent does not call any witnesses. In such a case the sequence of the hearing is varied in that the claimant's counsel is required to make his closing address before that of the respondent.

At the conclusion of the two closing addresses the arbitrator declares the hearing closed, and usually gives the parties an indication as to when his award will be published.

8 The award

SYNOPSIS

In making his award the arbitrator's primary objective is to define clearly, unambiguously, justly and enforceably, what the parties are to do and when they are to do it in order to resolve the matters in dispute. His secondary objective is to satisfy the parties, and in particular the losing party, that justice has been done. This chapter covers procedure for making the award: types, format and content of awards, and the manner in which the availability of an award is notified to the parties.

PROCEDURE

The arbitrator's award must be based on evidence *adduced at the hearing*. That evidence will in most cases comprise:

1. Documentary evidence contained in agreed bundles and referred to during the hearing
2. Oral evidence adduced by witnesses of fact and recorded in their proofs of evidence, in the arbitrator's notes (possibly supplemented by a transcription or a tape recording); and retained in the arbitrator's memory
3. Written reports and oral evidence of experts
4. Real evidence, if any, inspected before, during or after the hearing.

The arbitrator must, of course, also take account of the parties', or their counsels' or other representatives', submissions on questions both of fact and of law.

The source of the evidence adduced at the hearing, whether documentary, oral or real, is in general the parties' witnesses of fact and of opinion. There is, however, another possible source, namely the arbitrator himself:

and in dealing with evidence of that kind the arbitrator must exercise great care to ensure that it has been exposed at the hearing to any possible challenge by either party (see Chapter 7).

In *Fisher and Another* v. *P. G. Wellfair (1981)* (19 BLR 52) the Court of Appeal, Lord Denning MR (as he then was) presiding, unanimously dismissed an appeal against a High Court order removing an arbitrator and setting aside his award on the ground of misconduct. The arbitrator had used his special knowledge in forming a different view of the facts from that presented in evidence by expert witnesses, without providing an opportunity for the claimants to deal with that different view. In an *ex parte* hearing at which only the claimants' representatives were present the arbitrator had listened carefully to evidence of fact from two witnesses and to opinion evidence from four experts, and had intervened where he required clarification. But he had erred in failing to expose to the sole party present at the hearing an opinion he had formed from his own technical knowledge; namely that the cost of repairing a defective building was £13 000. The claimants' expert evidence had shown the amount of the claim to be £93 000, and in the absence of the respondent, that evidence was unchallenged at the hearing. Hence the claimant had no opportunity to deal with the evidence that the arbitrator had effectively taken from himself.

Should the arbitrator become aware, either during or after the hearing, of some aspect of the evidence adduced by the parties that is or appears to be inconsistent with his own knowledge and/or experience, his only satisfactory way of dealing with that matter is to invite the parties to comment on the evidence he has, in effect, taken from himself. If necessary he must reconvene the hearing for that purpose. Unless he does so he can only choose between making an award which he does not himself believe to be just, or making an award that is unfair in that it takes account of evidence that the parties have not had an opportunity to challenge. To choose the latter course could result, as in the *Fisher* case, in a successful application to the High Court under section 23 of the 1950 Act, for his removal on the ground of misconduct.

It is a part of the arbitrator's duty to determine matters both of fact and of law. Usually his first step is to deal with the facts; and where factual matters are in dispute, to determine from the evidence what are the true facts upon which to base his award. *Fact* in this context includes matters of opinion; and, for example, one of the questions of fact that often arise in civil engineering disputes is whether or not the physical conditions or artificial obstructions encountered by the contractor were such as might reasonably have been foreseen by an experienced contractor.

In coming to a decision as to the truth where there is a conflict of factual evidence the arbitrator must *weigh* each witness's evidence: that is to say, he must form an opinion as to which evidence is the more reliable. In doing so he may take into account a witness's demeanour, and whether or not his evidence has been shown under cross-examination to be consistent. Where questions arise as to the credibility of a witness one of the functions of counsel is to expose, by skilful questioning, any inconsistency there may be in that witness's evidence.

His next step is to determine questions of law in the light of the parties' contentions thereon and of his findings of fact. His decision on such questions must take account of relevant case law: and where the parties are represented by counsel it is a part of their duty to draw to his attention the authorities on which they rely. If the authorities appear to be in conflict the arbitrator must decide which is the more relevant to the instant case. He may do so:

(a) by relying upon his own understanding of the questions and of the submissions of the counsel before him; or
(b) by taking legal advice; or
(c) before making his award and with the agreement of one or both parties, by seeking a determination of the questions by the court, as preliminary questions of law, under section 2 of the 1979 Act.

Where the arbitrator decides to adopt either of the first two courses of action he does not close the door to a review of his decision by the court, since (unless an exclusion agreement has been entered into) either party will be able to apply, after the award has been made, for leave to appeal under section 1 of the 1979 Act. Such leave is, however, given only subject to compliance with stringent conditions (see Chapter 10).

Secondly, where both parties request that he takes advice from some other counsel, clearly the arbitrator should do so. It is generally helpful in such circumstances for the arbitrator to obtain, from the parties' counsel, an agreed statement of the questions of law to be determined, and their agreement as to the counsel to be consulted. The arbitrator must, however, be careful not to delegate his decision to that counsel: he must make up his own mind in the light of the advice he receives. If one party requests that he take legal advice but the other rejects that proposal the arbitrator must himself decide how best to deal with the question.

The third option – namely, determination of the question of law by the High Court under section 2 of the 1979 Act, before making his award – may not be available. The court will not entertain an application under section 2 unless it is satisfied that determination of the application might produce substantial savings in cost; and that the question of law is one in respect of which leave to appeal would be likely to be given under section 1 of the Act. The granting of leave to appeal under that section is also subject to compliance with strict criteria (see Chapter 10).

Finally, having determined both the relevant facts and the law to be applied to those facts, the arbitrator must decide on the validity of the claims and counterclaims, on the remedies to which either or both parties are entitled, and on the way in which he ought to exercise his discretion in awarding costs.

TYPES OF AWARD

Final awards

Unless otherwise stated, an award is deemed to be final, and it concludes the reference. Thereafter the arbitrator is *functus officio*; meaning that he has discharged his duty. It follows that thereafter he has no jurisdiction to deal with any question or difficulty that may arise from his award.

The 1950 Act does, however, imply that the arbitrator retains certain residual powers even after making his final award, as follows:

Under section 17 the arbitrator is empowered to correct in an award *any clerical mistake or error arising from any accidental slip or omission.*

Under section 18(1) the arbitrator is empowered *inter alia* to tax the costs of the reference (i.e. the parties' costs). Some arbitrators take the view that where they intend to tax costs of the reference they must retain their power to do so my making only an interim award until such time as the taxation has been completed. Adoption of that procedure has the disadvantage that the dispute never becomes finally concluded if, as is often the case, the parties are able to agree as to the amount of their costs. Since it is unlikely that the power to tax costs could be exercised until after all other matters – including the *award* of costs – had been dealt with, usually in a final award, it is submitted that the arbitrator impliedly retains his power to tax costs of the reference provided that he indicates, in his final award, his intention to do so. Without such an implication the section 18(1) provision for taxation by the arbitrator would in most cases be ineffective.

Under section 18(4) of the 1950 Act an arbitrator who has made no provision in his award with respect to the costs of the reference is not merely empowered; he is *required*, on the application of a party and after hearing any party who desires to be heard, to amend his award to make provision for the payment of such costs.

Similarly, under section 22 of the 1950 Act, the arbitrator may be *required*, where the High Court so orders, to reconsider matters in his award.

Interim awards

Provision is made in section 14 of the 1950 Act for interim awards, and these are often used: where disputes can conveniently be divided into stages; where the determination of preliminary issues may save the time and cost of a prolonged reference; or, more generally, where the arbitrator's award of costs is to be dealt with separately from the substantive issues.

A question may, for example, arise as to whether or not a claim is time-barred under the limitation acts: if the respondent's submission that it is so barred is found to be valid, then the claim becomes invalid and there is no need to proceed further. Similarly any other dispute as to the validity in principle of a claim may often be dealt with as a preliminary issue, obviating the need for further pursuit of the claim if the defence succeeds.

The matters referred to in an interim award are however determined finally therein: the word *interim* does not imply that those matters are subject to review, but only that the award does not determine all of the matters in dispute. Thus, for example, an interim award in which the respondent is found to be liable to the claimant in respect of certain items of the claim but not other items provides a final determination of those issues of liability: leaving issues of quantum to be dealt with in another award.

Performance awards

Although awards are usually made in monetary terms, there is a provision, in section 15 of the 1950 Act, for an arbitrator to order *specific performance*; that is, to order that a party shall perform certain specified works, or hand over goods or rights, other than matters related to land or to any interest in land. In construction disputes the provision may be used, for example, to deal with remedial works required to be performed by the contractor for the original work, so as to ensure uniformity of the finished work, or to ensure that no question arises as to responsibility for future defects.

In general a performance award should not be made where a monetary award would resolve the dispute in a satisfactory way, because of the danger that the manner of performance of the work may lead to a further dispute. However, where there is some special reason why the respondent should carry out remedial works, or where both parties so request, the arbitrator should make an *interim* award covering performance of the specified work; thereby retaining his power to deal with any subsequent dispute that may arise from its performance. The award should specify a reasonable period of time for completion of the work, if necessary making provision for extensions of that period on reasonable grounds. It should also cover provision for reasonable access for carrying out the work, the parties being entitled to apply to the arbitrator for an order where arrangements cannot be agreed. Finally, there should be provision for the claimant to report to the arbitrator as to whether or not the work has been performed in accordance with the order; for the arbitrator to inspect the work done in the event of a dispute as to its adequacy or quality; and there should be provision for dealing with any such further dispute; and for dealing with all remaining matters such as costs.

Consent awards

Where, as often happens, the parties negotiate a settlement of the dispute before it reaches a hearing or an award, it is desirable that the terms of the settlement should be incorporated in a *consent award*. The purpose of so doing is to define clearly the matters that have been so settled; to define responsibility for costs; to enable a party to take enforcement proceedings in the event that the other party fails to comply with the terms of the settlement; and formally to bring the arbitration to a conclusion.

FORMAT OF THE AWARD

Before the 1979 Act became law it was the general practice of arbitrators not to give reasons in their awards, because of the fear that such reasons might provide a basis for a challenge through the High Court on the ground of an error on the face of the award. By keeping the award as brief as possible the risk of such a challenge was minimized.

Such awards were, however, highly unsatisfactory to the parties, and particularly to the losing party, who might see no logical reason for the arbitrator's decision.

The 1979 Act, and the *Nema* rules (see p. 77) have resulted in a marked change of practice in making awards. It is now recognized by most arbitrators that the High Court will not interfere with an arbitrator's award unless there are strong reasons for doing so. Hence arbitrators may, and should, include reasons in their awards, and they may do so without giving grounds for an appeal provided that they take reasonable care to ensure that their awards are soundly based and carefully presented (see SD/26).

The main elements of a reasoned award are:

Headings
Recitals
Issues in dispute and the parties' contentions
Outline of events leading to the dispute
Arbitrator's findings of fact
Arbitrator's holdings on questions of law
Arbitrator's award and directions
Fit for counsel (where appropriate)
Signature, date and witness's signature

Headings

Awards are headed with the standard introduction *In the matter of the Arbitration Acts 1950–1979 and in the matter of an arbitration between. . . .* If the arbitration is being conducted under some procedural law other than English law the Acts referred to should be altered accordingly: for example, in Northern Ireland arbitrations are conducted under the Arbitration Act (Northern Ireland) 1937.

Thereafter the claimant and the respondent are identified so that, within the award, they may be referred to by those descriptions.

Finally, there is a heading *Award, Interim Award, First Interim Award, Final Award*, or whatever may be appropriate. Unless otherwise stated an award is deemed to be final: but it is suggested that where one or more interim awards have been made, the heading *Final Award* should be used in order to avoid possible misunderstanding.

Recitals

The word *Whereas* is used to introduce a series of numbered paragraphs wherein the arbitrator recites the series of events that have led to his making the award. These include:

1. Identification, main purpose and form of the contract from which the dispute arose
2. The existence within that contract of an arbitration agreement
3. The provisions therein for appointing the arbitrator
4. The occurrence of a dispute
5. The manner in which the arbitrator was appointed
6. The arbitrator's acceptance of the appointment
7. Agreement of the arbitrator's terms of appointment
8. Outline of the interlocutory proceedings
9. Date of commencement and duration of the hearing.

It may also be necessary to refer to any special rules or procedures adopted, or events before or during the hearing; failures of either party to comply with directions; or any other matters which may have a bearing on the arbitrator's award of costs. For example, where the parties agreed to a *documents only* procedure; where either or both parties requested a reasoned award (notwithstanding that the arbitrator would in any case have given reasons); or where the parties agreed to some special provision as to the award of costs: such facts should be recorded in the award.

Issues in dispute and parties' contentions

A brief statement of the issues and of the parties' contentions: for example a statement of the basis of the claimant's claim and of the respondent's rejection: may be included at this stage. It may, however, be better in some cases to deal with the issues and parties' contentions in relation thereto in the context of the narrative referred to below: especially where there is a large number of contentious issues.

Outline of the events leading to the dispute

A narrative of events leading to the dispute, although arguably not necessary to the parties, who are well aware of those events, is helpful to any other person who may have occasion to read the award – such as a judge in the High Court to which application is made for leave to appeal. Furthermore, such a narrative provides confirmation to the parties of the arbitrator's understanding of the issues in question, and of the events that are especially relevant in making the award.

In some cases it is desirable that the narrative be divided into stages: for example the pre-tender period; the period from invitation to submission and acceptance of the tender; and the construction period.

Arbitrator's findings of fact

The arbitrator's decisions on questions of fact in dispute between the parties should be stated. His reasons for coming to those decisions need not be stated, because such decisions are not open to challenge even if leave to appeal is given: the arbitrator being the sole judge of fact. However, some arbitrators prefer to explain how they have come to such decisions: possibly by a simple statement that they prefer the evidence of A to that of B. In particular, where a finding of fact may appear to be perverse, a reason for that finding may help the party adversely affected by it to accept that he has been treated justly.

Arbitrator's holdings on questions of law

Under this heading it is usual for the arbitrator to state his holdings on legal issues in contention and, where necessary, the basis on which he reached his conclusions. He may, for example, examine how disputed constructions of a clause of the contract would operate in hypothetical circumstances.

In an address to the Chartered Institute of Arbitrators in 1981 Lord Justice Donaldson, as he then was, stated that arbitrators are not required to analyse the law. They should set down what happened – that is, their findings as to facts – and state their opinions as to the law applicable to those facts.

Arbitrator's award and directions

The *operative part* of the award is usually headed *I award and direct that* or *I award and declare that*: the latter being the appropriate form for a declaratory award; where, for example, in an interim award the arbitrator determines issues of liability but not of quantum.

In the operative part of the award the arbitrator defines, firstly, what action is required by one or both parties in order to resolve the substantive issues in the dispute; which may of course involve one or more items of claim and one or more items of counterclaim. The basis of evaluation of the net sum payable by one party to the other will in most cases have been defined earlier in the award, so that the award and direction is usually in the form of a direction that *A shall pay to B, no later than fourteen days after the* *date of this my award, the sum of £x in full and final settlement of all claims and counterclaims referred to me herein, together with interest* thereon for the period from . . . until the date of this award, in the sum of £y.*

Secondly, the award may deal with costs of the reference: that is, the parties' costs. Where the arbitrator has decided to deal separately, in a final award, with costs, the usual reference to costs in his interim award on the substantive issues is: *the costs of the reference shall be borne by such party or parties as I may direct in my final award, which will be made after I have heard both parties' submissions as to costs.*

The usual format of the operative part of an award dealing with costs is: *A shall pay B's costs of the reference: such costs, if not agreed, to be taxed by me.*

Finally, the costs of the award – i.e. the arbitrator's own fees and expenses – are dealt with: the usual form of wording (in an interim award) being: *The costs of the award, which costs I hereby tax and settle in the sum of £x, shall be borne by such party or parties as I may direct in my final award after having heard the parties' submissions thereon.* Where the costs of the award are dealt with in a final award an appropriate form of wording is: *The costs of the award, which costs I hereby tax and settle in the sum of £x, shall be borne by the respondent: provided that, if such costs have already been paid to me by the claimant, the respondent shall within fourteen days of the date of this award reimburse to the claimant the sum so paid.*

The basis on which the arbitrator exercises his discretion to award and to tax costs under section 18 of the 1950 Act is dealt with in Chapter 9.

Fit for counsel

The marking of an award *Fit for counsel* indicates to a taxing master or to whoever may be responsible for taxing costs that counsels' fees should be allowed. Generally awards are so marked where both parties appoint counsel.

Where only one party appoints counsel the arbitrator must decide whether or not that appointment was necesary: and in so doing he should take into account whether or not there were complex issues of law on which submissions were needed: and whether or not there were conflicts in the evidence of fact of witnesses, that required skilful cross-examination in order to discover the truth.

A third situation that sometimes arises is where a party, having initially intended not to appoint counsel, finds it necessary to do so only in order that the level of its representation should equal that of the other party: and makes that reason clear at the time of the appointment. In such cases the arbitrator might validly take note of a submission that the costs of counsel ought to be borne in any event by the party who insisted, against the wishes of its opponent, to appoint counsel; but he would so order only if it

* See p. 65.

appeared to the arbitrator that the appointment of counsel was not necessary.

Where a question arises as to whether or not an award ought to be marked *Fit for counsel* and the arbitrator decides against so doing it is sometimes suggested that he should mark the award *Not fit for counsel*. That practice is however to be deprecated. A bald statement to that effect would be discourteous and unsatisfactory both to the counsel and to the appointing party. The usual, and preferred, practice is for the arbitrator merely to omit the statement *Fit for counsel* and to explain in the award his reasons for so doing.

Signature, date and witness's signature

The arbitrator should sign his award in the presence of a witness, who should also sign it and state his address and occupation.

INTEREST

Where the arbitrator finds that the sum or sums of money he awards in respect of claims or counterclaims ought to have been paid at some earlier date he may, and in general should, award interest on those sums; either, where the contract so specifies, under the contract itself, or under section 19A of the 1950 Act.

Where the contract makes provision for interest on late payments, and where that provision is valid, it should be used as thc basis for the interest awarded. Failing such provision the arbitrator should, unless there is some valid reason for not doing so, award interest under the statutory provision.

Under Clause 60(6) of the ICE Conditions, fifth edition, provision is made for the payment of interest at 2% over base rate on sums which the engineer fails to certify or the employer fails to pay. In *Morgan Grenfell (Local Authority Finance)* v. *Seven Seas Dredging (No. 2) (1990)* (51 BLR 85) His Honour Judge Newey QC, Official Referee, upheld the arbitrator's finding that the contractor was entitled to interest on the amount by which an engineer's certificate was increased by the arbitrator: and that the contractor is entitled to interest on the interest he should have received earlier.

The corresponding provision in the ICE Conditions, sixth edition, namely Clause 60(7), has been drafted so as to incorporate the effect of the court decision in the *Seven Seas* case; in that the clause expressly covers circumstances where an arbitrator finds that the engineer failed to certify, and provides that in such circumstances interest, at 2% over base rate, is to be compounded monthly.

The statutory provision, namely section 19A of the 1950 Act, gives the arbitrator a discretionary power to award *simple* interest at such rate as he thinks fit, for a period ending not later than the date of the award: and it also makes provision for such interest on sums which is *the subject of the reference but is paid before the award*. In exercising his discretion the arbitrator must, of course, do so judicially: and in general the appropriate rate of interest is, it is submitted, 2% over base rate.

Records of variations in base rate and tables of interest factors are given in Appendix G.

PUBLICATION OF THE AWARD

When it has been prepared, signed and witnessed, the award is said to have been *published*. That word does not, however, carry its usual meaning: only the parties are entitled to see the award.

The usual means whereby the arbitrator ensures that he will be paid for his services (usually, ultimately, by the party against which he has found) is to notify both parties that his award is available for collection by them or for dispatch to them on payment of his charges in the sum of £*x* (see SD/28). Upon receipt of payment the arbitrator dispatches the award to both parties. The award itself, where it deals with costs, makes provision for the likelihood that the party to whom costs are awarded will initially have paid the arbitrator's charges (see SD/27).

9 Costs

SYNOPSIS

Wide powers to award costs and to determine their amount are vested in the arbitrator under section 18 of the 1950 Act. But those powers must be used judicially: and where an unusual award of costs is made, reasons should be given.

1950 ACT PROVISIONS

Section 18 of the 1950 Act provides that costs of the reference and award are in the discretion of the arbitrator, who may direct who shall pay those costs and may determine their amount.

The costs of the reference are the parties' costs, covering their legal representation, experts, witnesses' expenses, and any other costs incurred from the date of commencement of the arbitration until the end of the hearing. Costs of the award are the arbitrator's costs, i.e. his fees and expenses. Thus the Act empowers the arbitrator to determine the amount of his own charges, but a safeguard against the abuse of this power is contained in section 19. Under that section an arbitrator who refuses to deliver his award except upon payment of his fees may be ordered to do so by the High Court, upon application by one of the parties and upon that party paying the sum demanded into court. The taxing master of the court thereupon taxes (that is, assesses) the amount of the arbitrator's fee, pays it to him and refunds the balance (if any) of the payment into court to the applicant. Where this procedure is used the arbitrator has the right to appear before the taxing master. A party which has agreed in writing to the arbitrator's fees is not entitled to use section 19 of the Act.

RESPONSIBILITY FOR COSTS

The basic rule used to determine who should pay the costs incurred in arbitration proceedings, as in the courts, is that 'costs follow the event' – meaning in general that the successful party should be awarded his costs. No difficulty arises in applying this rule in a simple case, where the claim either succeeds or fails in its entirety, and where there is no default on the part of the successful party.

In practice, however, complications often arise: for example a claim may succeed in part; or there may be a counterclaim which also succeeds wholly or in part; or the respondent may have made an offer which the claimant should have accepted; or costs may have been occasioned unnecessarily by the successful party. Arbitrators have often been criticized for their tendency to apportion costs in such cases; for example, to award one half of the claimant's costs where only one of two claims succeeds.

The arbitrator should, it is suggested, base his award of costs upon his decision as to which party's action or inaction resulted in the costs being incurred, and should award costs against that party to the extent that they were incurred necessarily. Hence a party whose claim succeeds in part should be awarded costs, but omitting costs in respect of time spent unnecessarily, in pursuit of obviously invalid claims. Where there is a claim and a counterclaim, both of which succeed in whole or in part, the net winner should be awarded his costs; but again to the extent that they were incurred necessarily. This follows from the fact that the net winner had to proceed with the arbitration in order to obtain the net sum to which he was found to be entitled.

This general rule for the award of costs in a claim/counterclaim situation derives from the judgment of Mr Justice Donaldson (as he then was) in *Tramountana Armadora* v. *Atlantic Shipping Co. (1978)* (1 Lloyd's Rep 391). In that judgment Lord Donaldson set down with great clarity the principles to be applied in the award of costs, and he drew attention to several inadequacies of the law as it then existed; some of which have since been remedied. Besides giving the basic rule that the nett winner in a claim/counterclaim situation should be awarded costs, Lord Donaldson referred *obiter* to the position where the claim and counterclaim have separate existence, and where it may be appropriate for the arbitrator to award costs separately for the claim and for the counterclaim. That situation is however unlikely to arise from construction contract disputes, in which counterclaims, where they are pleaded, almost invariably arise as part of the defence to a claim.

The judgment in the *Tramountana* also emphasizes that the discretion as to the award of costs lies with the arbitrator and not with the court. While the court would give directions as to the principles to be followed by the arbitrator in the exercise of his discretion it would not and could not exercise that discretion on the arbitrator's behalf.

Where one party makes an offer to settle the claims in a sum sufficient to cover the amount ultimately awarded by the arbitrator, plus costs incurred by the claimant up to the date of the offer, then the responsibility for the continuation of the dispute rests with the offeree, who should be made to bear costs incurred after the date of the offer.

Where a successful party is responsible for unnecessary costs – for example by failing to appear at meetings which have to be aborted, or by making late applications to amend pleadings, thereby increasing the costs of the hearing – that party should be made to bear all such unnecessary costs. This requirement is usually implemented by means of an order by the arbitrator, made at the time of the default, to the effect that the costs of the aborted meeting and of the arbitrator's order shall be borne by the defaulting party 'in any event' – that is, irrespective of the outcome of the arbitration.

Whenever an arbitrator makes an unusual award of costs – i.e. one in which the rule *costs follow the event* is not applied either to the whole or to a part of the costs – his reasons for doing so should be stated in the award, whether or not the award includes reasons for the substantive matters determined therein. Such reasons might be stated as 'having regard to the offer made by the respondent . . .' or 'having regard to the failure of the claimant to comply with my directions to attend a meeting on . . .' or some such brief explanation of what might otherwise appear to be an incorrect award of costs.

OFFERS TO SETTLE

It will be seen from the above that the making of an offer of sufficient magnitude to warrant its acceptance by the offeree protects the offeror against costs as from the date of the offer. The machinery whereby such offers may be made requires consideration.

Open offers

Either party may make an offer to settle to the other, by way of an 'open' letter (that is, a letter not marked 'without prejudice') which may therefore be produced in evidence. The disadvantage of this procedure to the offeror is that the letter might be interpreted by the arbitrator as an admission of liability notwithstanding that the offeror intends to contest liability if the arbitration proceeds.

Sealed offers (calderbank letters)

A calderbank letter* is an attempt to provide in arbitration a procedure corresponding to a payment into court in litigation. Under the court

* See SD/24.

procedure, a defendant may pay into court a sum representing his offer in settlement of the claim against him. The plaintiff is notified of the offer and if he accepts it he is awarded costs. If he rejects the offer the action proceeds. The fact of the payment into court is not made known to the judge hearing the case until after he has given judgment on the claim, but before he deals with costs. In this way the offer cannot prejudice the action by any implication of admitted liability, but it can be taken into account as a possible factor affecting the award of costs.

In arbitration it is not possible to use this procedure because there is no court into which a respondent can make a payment. A practice has therefore been developed under which the respondent makes an offer to the claimant in writing, which is protected from discovery by being marked *without prejudice save as to costs*. Such a communication is known as a *calderbank letter*. If the offer is not accepted a copy of it in a sealed envelope is handed to the arbitrator at the conclusion of the hearing, with a request that the envelope is not opened until after he has made his substantive award, but before he deals with costs.

The defects of this procedure are, firstly, that although the arbitrator is unaware of the amount of the offer when making his substantive award he need not be clairvoyant to know that the envelope contains a copy of an offer. Secondly, the offer in writing is not equivalent to a payment into court because there can be no guarantee that funds were available, and would have been paid, had the offer been accepted.

The first of these defects may be overcome by the interim award procedure referred to in the following section.

Until recently there existed some doubt as to how an offer should be marked. That doubt has been resolved by the Rules of the Supreme Court (Amendment) 1986 (SI 1986 No. 632), which in paragraph 5 amends Order 22 Rule 13 of the RSC by providing for offers to be marked 'without prejudice save as to costs'. Such marking ensures that the existence of the offer is not made known to the court until the question of costs falls to be decided. The rule also provides that the procedure is not effective in relation to costs if the party making the offer could have protected his position by making a payment into court.

Interim awards

The problem of dealing with offers arises in arbitration because, except in administered arbitrations, there is no 'court' into which payment may be made, or which can be made aware of the existence of the offer without its coming to the notice of the arbitrator. It is, however, possible for the arbitrator to notify to the parties his intention to make his substantive award in the form of an interim award, after publication of which he will hear the parties' representatives on the question of costs. In an important reference, or in any reference in which it appears likely that an offer may have been made, the arbitrator would be wise to adopt this procedure. It would of course be open to him to comply with their wishes if both parties request that the substantive award and costs are dealt with together. Where neither party makes such a request the arbitrator would not necessarily be correct in inferring that an offer had been made; it could be that the parties are unaware of the reason underlying his suggestion.

A party wishing to adopt this procedure, in a reference in which the arbitrator does not himself suggest it, could ask the arbitrator for an opportunity to address him on costs after his substantive award is published. Such a request is unlikely to be refused; indeed to do so might well provide grounds upon which the party making the request could make a successful application under section 22 of the 1950 Act for remission of the award of costs to the arbitrator for reconsideration.

After having heard the parties' addresses on costs the arbitrator incorporates his orders as to costs in a final award, which is published in the usual way.

FAILURE TO AWARD COSTS

Provision is made in section 18(4) of the 1950 Act whereby a party may, within 14 days of the publication of an award that fails to deal with costs of the reference, apply to the arbitrator for him to make an order directing by and to whom those costs shall be paid, and requiring the arbitrator, after hearing any party which may wish to be heard, to amend his award by adding directions as to costs. The procedure defined in this subsection appears to suit the requirements admirably, but strangely it is rarely used; possibly because an arbitrator who failed to deal with costs in his award might be thought to have done so inadvertently.

TAXATION OF COSTS

'Taxation' in relation to costs means 'determination of amount': the term is not related in any way to taxes imposed by Inland Revenue or Customs & Excise. The arbitrator is empowered under section 18(1) of the 1950 Act to 'tax or settle' the costs of the reference and the costs of the award.

The practice of most arbitrators is to tax the costs of the award: i.e. their own costs. This is because arbitrators know how much time they have spent in dealing with, and have allocated to, the reference: they know what hourly and daily rates are usual or have been agreed with the parties and they know the amount of their expenses. Hence it would be absurd, and an unnecessary waste of time, for the costs of the award to be left to be taxed by the High Court, which is the alternative to the arbitrator's exercise of his own power.

Until the last few years many arbitrators have left the taxation of costs of the reference – the parties' costs – to the court. The reason for so doing appears to have been the belief that taxation is necessarily a complex and highly technical procedure requiring many years of experience in a highly specialized field. That reason does not, however, appear to have deterred the arbitrator from taxation of his own costs.

More recently the benefits of taxation of the costs of the reference by the arbitrator, where those costs are not agreed between the parties, have been gaining recognition. Unlike a taxing master in the High Court, the arbitrator is already fully aware of the facts of the case: he does not need to spend time in ascertaining those facts. He is already aware of the contribution if any made by any experts who may have been appointed, so that he is able to form an opinion as to whether or not experts' costs were incurred necessarily. Thirdly, he is likely to be able to undertake the taxation, if it becomes necessary, with much less delay than would be incurred in the court.

For these reasons it is submitted that the arbitrator should, either at the preliminary meeting or at the hearing, offer to undertake taxation of the costs of the reference if those costs are not agreed. In the author's experience that offer is invariably accepted. However, if one party accepts and the other rejects, the arbitrator might take account of the reason for the rejection: if, for example, it appears that the sole or the main objective is to incur delay in bringing the reference to a conclusion then the arbitrator might well decide to exercise his power notwithstanding an objection by one of the parties.

Where the parties agree, or the arbitrator decides, that he will use his power he should so state in his award (see, for example, SD/27). He should then, if a difference arises as to taxation, ascertain whether or not the parties agree to this taxation being undertaken upon receipt of written representations only, or if the parties require the opportunity of an oral hearing. In either case he should take note of the successful party's bill of costs, of the losing party's objections thereto and the successful party's reply, before coming to his decision.

Whether or not the arbitrator undertakes taxation he should always determine the scale of costs awarded. Section 18(1) of the 1950 Act refers to 'costs to be paid as between solicitor and client': being one of the scales that were in use at the date of the Act. There have since been several changes, the current scales being those given in the Rules of the Supreme Court (Amendment) 1986 and defined as follows:

(a) *Standard basis*

'. . . a reasonable amount in respect of all costs reasonably incurred and any doubts which the taxing officer may have as to whether the costs were reasonably incurred or were reasonable in amount shall be resolved in favour of the paying party'.

(b) *Indemnity basis*

'. . . all costs . . . except insofar as they are of unreasonable amount or have been unreasonably incurred and any doubts which the taxing officer may have as to whether the costs were reasonably incurred or were reasonable in amount shall be resolved in favour of the receiving party'.

The distinction may not be immediately apparent: however, such a distinction does exist and is important. The burden of proof that the costs were both reasonably incurred and reasonable in amount is, in the case of costs on the standard basis, borne by the receiving party; while in the case of costs on the indemnity basis the paying party bears the burden of proof that costs were not reasonably incurred or not reasonable in amount.

Where no scale is stated the standard basis is deemed to apply. In general the arbitrator should award costs on the standard basis, and should exercise his power to award on the indemnity basis only where some good and legally valid reason exists for using the higher scale.

AVOIDING UNNECESSARY TAXATION

The arbitrator should recognize that taxation of costs can itself involve costs, and may be protracted. It follows that an award that obviates or minimizes the need for taxation may be more sensible, and may do more good to the parties, than one based upon strict principles. For example, where an offer deemed by the arbitrator to be acceptable was made at a date by which half of the total of costs had been incurred, strict application of the rules would lead to an award that the respondent bears the claimant's costs up to the date of the offer, and the claimant bears the respondent's costs thereafter. An order that each party bears its own costs, although not strictly accurate, is likely to be more beneficial to both parties in that it obviates the need for taxation.

EFFECT OF TAXATION

One of the less equitable aspects of the law in relation to taxation of costs is that even where a party is found to be blameless in an arbitration, or in litigation, and is awarded costs, the sum allowed on taxation is almost invariably less – and often substantially less – than the costs actually incurred by that party. It is not within the power of the party in question, or the arbitrator, to correct this injustice. The party can, however, to some extent mitigate its loss by ensuring as far as possible that its costs fall within the definition of standard basis costs; namely that they are reasonably incurred and reasonable in amount.

The changes introduced under the Rules of the Supreme Court (Amendment) 1986 referred to above should in general ensure that the successful party is awarded a higher proportion of his costs on taxation than was the case under earlier rules.

COSTS OF THE AWARD

Whatever decision is reached by the arbitrator in relation to the parties' costs should apply equally to the arbitrator's costs: that is, the costs of the award. Thus the party found to be responsible for the other party's costs should also have to bear the arbitrator's costs. Where responsibility is allocated to one party up to a certain date and thereafter to the other, the arbitrator's costs for each period should be determined and awarded accordingly.

10 Finality of the award; enforcement; appeals

SYNOPSIS

Although the objective of arbitration is to produce an award which is just, final and enforceable, there are reasons why this objective is not always achieved. There may be accidental errors in, or omissions from, the award; it may have been procured improperly or it may contain errors of law. This chapter covers the enforcement of valid awards and the means by which erroneous awards may be rectified or nullified.

ENFORCEMENT

Where the losing party fails to honour an award, application may be made by the other party under section 26 of the 1950 Act to the High Court for judgment in the terms of the award. Provided that the court is satisfied that the award is valid judgment will normally be given, and the means of enforcement thereafter is similar to that of any other judgment of the High Court.

CORRECTION OF ACCIDENTAL ERRORS

Although the arbitrator, having made and published his award, is *functus officio* – meaning that he has discharged his duty and therefore that his authority as arbitrator is ended – he is empowered under section 17 of the 1950 Act to correct 'any clerical mistake or error arising from any accidental slip or omission'. It is rarely necessary in practice to invoke this power, because arbitrators should, and usually do, take great care to ensure that their awards do not contain mistakes or errors. The power given under this section cannot, of course, be used to correct other types of error, such as errors of law. In *Mutual Shipping Corporation* v. *Bayshore Shipping Co. (1985)* (TLR, 14 January 1985) it was held in the Court of Appeal that an arbitrator who mistakenly attributed evidence to the wrong party, causing him to make an award in favour of the wrong party, had power to correct that error under section 17 of the 1950 Act.

AMENDMENTS TO DEAL WITH COSTS

Section 18(4) of the 1950 Act covers the situation where an award fails to deal with the costs of the reference. This provision is dealt with under Costs (p. 71).

CORRECTION OF ERRORS OF LAW

An arbitrator is judge of matters of fact and of matters of law. 'Fact' in this context includes opinion on technical matters but not on questions of law; 'law' covers the interpretation (or, in legal terminology, the 'construction') of a contract. The arbitrator's findings of fact are final in that they are not subject to any right of appeal, but his decisions on questions of law may, in certain circumstances, form the subject of an appeal to the High Court.

It is, however, open to both parties to enter into an agreement, after the dispute has arisen, excluding their right of appeal. Such an agreement is recognized under section 3 of the 1979 Act and is termed an 'exclusion agreement'.* Its effect is to make the arbitrator sole judge of questions of law; and where the main concern of the parties is to reach finality an exclusion agreement, entered into at the commencement of the arbitration, will ensure that that aim is achieved.

Where no such agreement is in operation at the time of publication of the award, section 1 of the 1979 Act (which applies to all arbitrations commenced on or after 1 August 1979 and to other arbitrations where the parties agree that it should apply) provides a limited right of appeal from the arbitrator's decisions on questions of law. The intention of the Act, however, is to obviate what had become almost standard practice prior to its enactment, namely the pursuit of sometimes spurious 'points of law' through the hierarchy of the courts by means of the 'special case' procedure, which now no longer exists. The court will accordingly require to be satisfied that the question of law is a genuine and an important one, and may impose conditions upon the granting of leave to appeal.

A party wishing to appeal under this section of the 1979 Act must either obtain the consent of the other part to the reference, or must be given leave by the court. In most cases leave to appeal will be sought by a claimant whose claim has been disallowed, or by a respondent who has been found liable to the claimant; in both of these situations the successful party will

* See SD/14 for example.

almost certainly oppose any application to appeal. Hence, in general, the party wishing to appeal will have to apply for leave from the court.

In dealing with such applications the High Court requires to be satisfied that determination of the question of law could substantially affect the rights of one or more parties to the arbitration. Leave to appeal, if granted, may be made subject to the applicant complying with such conditions as may be considered by the court to be appropriate. It is to be expected in general that such conditions will include the payment into court of any sum directed in the award to be paid; thereby ensuring that the applicant does not gain an advantage from postponement of the day of payment.

In *BTP Tioxide* v. *Pioneer Shipping (The Nema) (1981)* (3 WLR 292), the arbitrator's award had been reversed in the High Court, reinstated in the Court of Appeal, and upheld in the House of Lords. The late Lord Diplock said in the House of Lords: 'It is not self-evident that an arbitrator . . . chosen by the parties for his . . . experience and knowledge of the commercial background and usages of the trade in which the dispute arises is less competent to ascertain the mutual intentions of the parties than a judge of the Commercial Court, a Court of Appeal of three Lords Justices or even an Appellate Committee of five Lords of Appeal in Ordinary.' Hence he inferred that the parties to an arbitration agreement, having chosen that tribunal, are content to take the risk that an arbitrator will make mistakes of law. Lord Diplock went on to say that that inference did not mean that the parties had agreed to accept an award founded upon a manifest error of law: and he made a distinction between disputes over contracts in standard terms and those which are 'one-off' clauses or situations.

The Nema judgment gave rise to a degree of uncertainty as to how the guidelines should be applied, mainly in shipping disputes, which provide the main body of case law in matters relating to arbitration. After a series of applications under the 1979 Act a further case, *Maritime Transport Overseas* v. *Unitramp Salem Rederierna, The Antaios (1981)* (2 Lloyd's Rep 284), resulted in a further appeal to the House of Lords. The decisions in the two cases so appealed, known as *The Nema* and *The Antaios*, provide the rules by which applications for leave to appeal under section 1 of the 1979 Act are determined: which rules may be summarized as follows:

1. *Where the question of law involved is the construction of a 'one-off' clause in a contract, leave to appeal should not be given unless it is apparent from a mere perusal of the award that the meaning ascribed to the clause by the arbitrator is obviously wrong.*
2. *Where the question of law involved is the construction of a standard form of contract leave to appeal should not be given unless a strong* prima facie *case has been made out that the arbitrator had been wrong in his construction.*

Equally stringent conditions are imposed by the 1979 Act and by case law on the granting of leave to appeal from decisions of the High Court.

REASONED AWARDS

The practicalities of an appeal under the 1979 Act are closely related to the giving of reasons for the award, because where no reasons are given it may not be possible to determine whether or not the arbitrator has made an error of law. In its shortest form an award might consist of recitals followed by a direction as to payment by one party to the other: and unless there is some obvious error in such an award (for example, where a respondent has admitted liability for a sum greater than that awarded) such an award could not be challenged under this section of the 1979 Act.

One of the stated intentions of that Act is to encourage arbitrators to give reasons; but they are not required to do so. Furthermore, if both parties request that reasons be not given, then the arbitrator should respect that request. If reasons are required by one or both parties the arbitrator should usually comply, although he is not obliged to do so. Such a requirement, notified to the arbitrator before publication of his award, does however ensure that adequate reasons can be made available to the High Court if they are required in connection with an appeal.

Where the parties are content to rely upon the arbitrator's own decisions upon questions of law they may request that he does not give reasons in his award: alternatively, or additionally, they may enter into an exclusion agreement. Where rights of appeal are to be preserved the party or parties wishing to preserve them should require that reasons be given. Unless reasons are required by one of the parties before the award is made, or unless there are special reasons why reasons were not required, the High Court cannot order the arbitrator to give reasons (section 1(6) of the 1979 Act).

Appeals under section 1 of the 1979 Act arising from construction contracts are dealt with in the Official Referees' Courts. Notice of appeal must be served within 21 days after publication of the award (RSC Order 73 Rule 5). Where leave to appeal is given the options open to the High Court on determination are to confirm, to vary or to set aside the award, or to remit it to the arbitrator for reconsideration, with the court's opinion on the question of law which was the subject of the appeal.

MISCONDUCT BY THE ARBITRATOR

Section 23 of the 1950 Act empowers the High Court to remove an arbitrator who misconducts himself or the proceedings. In such cases, or where it has been improperly procured, the award may be set aside by the court. A less severe remedy is available under section 22 of the 1950 Act, which empowers the court to remit the matters referred, or any of them, to the arbitrator for reconsideration.

Where the award is set aside all of the proceedings that have led to it are null and void. The parties, if they wish to proceed with the arbitration, must appoint a new arbitrator and recommence the proceedings: hence many of the costs incurred in the aborted reference are likely to be repeated. On the other hand remission to the arbitrator for reconsideration is unlikely to result in any substantial delay or additional cost.

'Misconduct' in this context does not necessarily imply, although it includes, immoral behaviour. The use of the term is often criticized because it fails to distinguish between behaviour which is morally reprehensible and that which is merely mistaken, such as might result from inexperience. Examples of behaviour in the first category include the acceptance of bribes or other inducements to show favour to a party, corruption, fraud or dishonesty, or any form of bias induced by the arbitrator's relationship with or interest in a party. In the sense of mistaken behaviour misconduct includes unintentional failure to disclose a relationship with a party which is unlikely to influence the arbitrator's award, improper delegation of his duty, hearing inadmissible evidence or refusing to hear admissible evidence, hearing evidence in the absence of a party (except where the rules relating to *ex parte* proceedings have been observed), hearing or receiving any submission or communication from one party without the knowledge of the other, refusal to allow time for an application to the court on any matter in respect of which the parties have a right to apply, or the use of the arbitrator's expert or factual knowledge of matters at issue without advising the paarties of his intention to do so or giving them an opportunity to challenge that knowledge.

One of the matters often misunderstood by arbitrators is the extent to which they should use their own technical knowledge in reaching their decision on the issues they are called upon to determine. In disputes arising in the construction industry arbitration agreements incorporated in the relevant forms of contract provide for the arbitrator to be appointed either by agreement or by the president of the appropriate professional body. Thus the parties make provision, in their agreement, for the appointment of an arbitrator having relevant technical knowledge; and in doing so they impliedly agree that the arbitrator should use that knowledge.

An arbitrator who is so chosen for his knowledge and experience not only may, but *ought* to use that knowledge and experience in coming to his decision. He must, however, ensure that the result of his so doing – the evidence which he effectively takes from himself – is exposed to the parties, so that they may comment on, or challenge, that evidence in the same way as they might challenge evidence adduced by the opposing party. (See *Fisher and Another* v. *P. G. Wellfair (1981)* (19 BLR 52); p. 57.)

In both of its forms misconduct covers any action contrary to the principles of natural justice, which require that no man may be a judge in his own cause, and that every party has a right to be heard and to challenge any statement or document prejudicial to its case.

Application under section 23 of the 1950 Act for setting aside an award on the ground that it has been improperly procured is unlikely, because any allegation of illegality of the contract or of invalidity of the arbitration agreement in a contract is usually taken long before the arbitration reaches an award. Nevertheless the power is there for use in appropriate situations.

Where application is made under section 23 of the 1950 Act for removal of an arbitrator and for setting aside his award it is for the court to decide whether or not to allow the application, or to remit the award under section 22. Generally the latter power is used where the matters referred are of a minor nature and as such are capable of being remedied by the arbitrator on having the defects pointed out to him. It is, however, recognized that the arbitrator, having committed himself to a view of the matters referred, may find it difficult to approach any question of that view with an entirely free mind. In such cases the court may feel obliged to adopt the more serious remedy of setting aside.

Matters which might be remitted include decisions upon interlocutory matters, such as granting or refusing applications for extensions of time or for the production of documents for which privilege is claimed, or defects in the award such as failure to deal with all of the matters referred to the arbitrator, or the inclusion of matters outside the reference.

Applications under section 22 or 23 of the 1950 Act must be made within 21 days of the date of publication of the award (RSC Amdt 3 1979: see p. 191). Where matters are remitted by the court to the arbitrator for reconsideration he must make his award within three months of the order, unless the order provides otherwise.

11 The contractor as claimant

SYNOPSIS

Probably the most usual form of a contractor's involvement in arbitration is as claimant, where a dispute arises from rejection of his claims. This chapter examines the origin of such disputes and ways in which they may sometimes be avoided. It also seeks to define the action that should be taken by a competent contractor: firstly, to minimize the likelihood of his having to invoke arbitration; and, secondly, where it is unavoidable, to ensure that the case he presents is valid and convincing. That action commences with the preparation of the tender.

THE TENDER

A contract is created by a valid acceptance of a valid offer. Certain other elements must also be present: there must be consideration, an intention to create a legally binding relationship, and certainty; but these are usually adequately covered in the offer, which in construction parlance is the tender.

Whether or not the tender documents issued by the employer so provide, the tenderer may limit the validity of his tender to a stated period of time. If by so doing he contravenes a requirement of the employer, then the courses open to the employer are to waive that requirement, to negotiate for removal of the tenderer's time limit, or to ignore the tender. In any event, and notwithstanding any undertaking he may have given, the tenderer may at any time *prior to its acceptance* withdraw the tender.

If the tenderer neither limits the period of validity of his tender nor withdraws it, then it remains open for acceptance with a 'reasonable' period of time: the question 'what is reasonable?' being one for decision by the courts in the event of a dispute.

In order to safeguard himself the tenderer should withdraw any tender that has not automatically lapsed through a time limit, when he no longer wishes it to be accepted: otherwise he may receive an unwelcome reminder, in the form of an acceptance, that the tender is still valid. If the tenderer wishes to challenge the validity of the acceptance on the ground that it has not been issued within a reasonable time he should give notice to that effect immediately, and should not take any step that might imply his acceptance of the existence of a contract. It is too late, at this stage, to demand an increase in the tender sum to take account of inflation, although such an increase could possibly be negotiated with an employer who accepts the tenderer's contention that because of the passage of time the tender has lapsed. If the tenderer merely accepts the existence of the contract created by acceptance, however late, of his tender, then it is useless to claim at some later date that he is entitled to compensation in respect of the delay, unless such a provision is already built into the contract as a 'variation of price' clause.

In preparing the tender the contractor should remember that in the event of a dispute arising, for example from a varied quantity or type of work, the basis upon which the work included in the tender was priced will almost certainly be relevant, and pricing notes may have to be produced on discovery (see p. 36). For this reason the estimator's notes should be clear, logical and legible: especially where, for example, the make-up of an item includes a fixed sum in addition to a rate related to the quantity. In such a case the contractor is entitled to, and must be able to justify, an enhanced rate should the quantity be reduced. Conversely, where the pricing of an item depends upon the availability of a limited quantity of material, such as filling, at a cheap price, an *increase* in the quantity required by the employer may justify an enhanced rate. The apparent illogicality of the contractor's seeking an enhanced rate in both of these instances of variation may not be readily acceptable to the employer or to his engineer: it is only by making available pricing notes and evidence in support of them, such as quotations, that valid claims of this nature may be established.

Again, the contractor should ensure that his methods and construction programme are clearly defined and are available for reference in the event of a dispute. In some cases it may be desirable, where the construction methods or programme have an important effect on the amount of the tender, to include the method statement or programme as part of the tender documents: otherwise the contractor could find himself in the position of having his tender accepted and his method or programme later disapproved.

This situation may arise where a tenderer wishes to complete the work in less than the time allowed for it in the tender documents, in order to reduce his oncosts. In *Glenlion Construction* v. *The Guinness Trust (1987)* (39 BLR 89), a building contract on the JCT Form included a special provision analogous to that of Clause 14 of the ICE Conditions. The contractor submitted a programme in accordance with that provision, showing completion before the contractual date for completion. It was held by His Honour Judge Fox-Andrews QC, Official Referee:

1. that the programme did not relieve the contractor of his responsibility to complete by the contractual date for completion,
2. that insofar as the programme showed completion before the contractual date for completion, the contractor was entitled to carry out the works in accordance with the programme, and
3. that there was no implied term of the contract that, if and insofar as the programme showed completion before the contractual date for completion, the employer and his agents and servants should so perform the contract as to enable the contractor to carry out the works in accordance with the programme.

Where a tender is based on a shorter time for completion than that specified in the tender documents a contractor wishing to ensure that he is entitled to the employer's and the engineer's collaboration in achieving that aim should make his tender conditional upon an appropriate reduction in the contractual time for completion. A corollary of such a condition is that the contractor incurs an *obligation* to complete within the reduced time for completion, and becomes liable for liquidated damages if he fails to do so; subject, of course, to the contractual provision for extension of time.

Where the tenderer finds it necessary to qualify his tender he should ensure that the qualification is clearly defined and that it is *incorporated in the tender*. A tenderer who submits a tender on the form provided, and submits qualifications in a separate covering letter to which no reference is made on the form of tender, runs the risk that the tender may be accepted as a separate document, and the covering letter ignored.

THE ACCEPTANCE

The other main element needed to create a contract is a valid acceptance: that is, an acceptance given during the period of validity of the offer, and in terms compatible with that offer. A simple statement 'I hereby accept your offer' is sufficient, provided of course that there is enough information to identify the offer to which it relates. If the 'acceptance' is not in terms compatible with the offer then it is not an acceptance at all, but a counter-offer, which requires acceptance by the other party before a contract can come into existence.

For example the acceptance of a qualified tender, subject to the removal of the qualifications, is a counter-offer which the tenderer may or may not accept. Alternatively he may possibly agree to remove the qualifications subject to an increase in the amount of the tender; thereby making a further counter-offer, which requires acceptance by the employer if a contract is to be formed. In this way there may be a series of offers and counter-offers: and it is only when the terms of an acceptance are compatible with the last counter-offer that a contract comes into being. There is, however, a danger that a party to such a series of counter-offers may find itself deemed in law to have accepted the last offer made. This may happen where a tenderer, having received a counter-offer, starts work on the site: such an action usually constitutes an implied acceptance of the last offer made: and even if the contractor makes it clear, before starting, that he is doing so pending agreement of the outstanding issues, such action is unlikely to help his contentions. For this amounts to an agreement to agree, which for obvious reasons is not enforceable in law.

Similarly, an employer who instructs a contractor to start work is likely to be deemed to have accepted the tender and so to have created a contract.

LETTERS OF INTENT

The usual objective of a letter of intent is to allow preliminary work to proceed while formalities such as financial or planning approvals are obtained. Where preliminary work is carried out in reliance on such a letter, and a contract later comes into being, no problem arises. Where for some reason the contract does not come into being differences often arise: usually from a failure of the parties to define, in the letter of intent, their rights and obligations in the event that that situation arises. Another possible source of difference is a contention by the contractor that the letter, either by itself or in conjunction with the employer's or the engineer's written or oral instructions, constituted an implied acceptance of the tender.

A draft letter of intent, based on the assumption that the contractor is to be entitled to remuneration on a *quantum meruit* basis for carrying out authorized work in accordance with the letter, is appended as SD/29.

THE CONTRACT

'Unless and until a formal Agreement is prepared and executed this Tender, together with your written acceptance thereof, shall constitute a binding Contract between us.'

The quotation, from the Form of Tender included in the ICE Conditions of Contract, merely sets out what is in any case the position in law. It follows that the question whether or not a formal agreement should be entered into and signed – and possibly sealed – by both parties is of little significance, since a perfectly adequate contract exists from the time of acceptance of the contractor's tender. The parties should, however, ensure that, where a formal agreement is made, it incorporates everything that has been agreed by the parties during any negotiations that may have taken place between the submission of the tender and the letter of acceptance. This it may do either by reference to letters between the parties setting out the agreed terms, or by a separate statement of the points that have been agreed during the negotiations. In any case the formal agreement, if made, should have annexed to it the tender drawings and any other information such as site investigation records made available to tenderers, the tender and all supporting documents, and the letter of acceptance.

The contract, whether in the form of the original tender and the letter of acceptance with appendices to those documents, or whether as a formal agreement, defines what is to be constructed and when it is to be completed: how much money is to be paid by the employer to the contractor, and the rights and obligations of each party in relation to the other. As such it is of vital importance in the determination of any dispute that may later arise, because the origin of any such dispute must lie in an allegation that some change has been made: either in the nature and/or quantity of the work to be performed, or in the way the contractor has performed it.

THE CONSTRUCTION PERIOD

Probably the most important single factor in ensuring that the contractor does not suffer unnecessarily from unforeseen difficulties, delays or alterations ordered during the course of the work is his maintenance of adequate records during the period of construction. The nature of civil engineering work is such that in almost every contract alterations in the work or in the manner or sequence of its construction must be expected. Ground conditions may not be as foreseen or as indicated by the borings and trial holes; weather conditions may be unfavourable; and alterations to the work may be needed in order to cope with the unexpected, to accommodate changes in the employer's requirements, to take advantage of changing technologies, or to correct errors in the design. All of these factors, together with the engineer's actions or omissions in his administration of the contract work, may give rise to variations, delays or disruptions of the work: and it is only by maintaining full and accurate records of these matters and events as they occur that the contractor can hope to obtain proper recompense for their effect on the costs he incurs.

In most civil engineering contracts the contractor's records should include:

(a) site diaries, in which are recorded the weather, the state of progress on each item of work, and the causes of any delays being encountered in any part of the work;
(b) progress charts, which in the case of underground work may include records of subsoil strata, water levels and so on;
(c) programmes of future work, including a dated copy of each revision of the programme;
(d) details of all plant and labour employed on site, and of their allocation to various parts of the project, including cost information;

(e) details of site oncosts: supervisory and non-productive labour, office costs including telephone, heating, cleaning, etc.; site workshops and mess-rooms; sanitation; water supply; site transport; small tools; insurances; watching and lighting;
(f) records of lump-sum costs incurred in establishing and in dismantling the site, such as costs of transporting plant to and from the site, erection of offices and sheds and constructing access roads;
(g) all correspondence with the employer, engineer and engineer's representative – this should include variations, instructions, amended drawings; the contractor's confirmation of oral instructions given by the engineer or by his representative; and the contractor's notifications to the engineer of information required or of any other matter causing or likely to cause delay to the progress of the works;
(h) correspondence with other parties, such as subcontractors, suppliers and any other organizations or persons who may be involved in the provision of goods or services required in connection with the works.

It should be recognized that the major element of most claims arising from civil engineering contracts is the cost of labour, plant and site oncosts during periods of delay. Hence it is important to the contractor that he should be able to produce evidence as to the causes of any delays suffered, of his having notified the engineer of those causes, and the effects of the delays in terms of additional costs. He should also ensure, as far as may be practicable, that notices are served as laid down in the contract. Failure to serve such notices may not be fatal to claims, but may well give grounds for a contention by the employer that he was thereby precluded from taking avoiding action, or that he was unable to check and record costs resulting from the delay.

CLAIMS

Where it appears to the contractor that he is entitled to be paid some additional money in respect of additional costs incurred by him, he should set out the facts giving rise to the claim, identify the clause of the contract entitling him to payment, and calculate the amount of the claim, including where appropriate allowances for oncosts and for profit. In general such a claim should be lodged as soon as possible after the events from which it originates but, where necessary, information should be given piecemeal, as it becomes available. For example, where the whole or any part of the works is held up because of lack of information, notice should be given of the need for such information as soon as the contractor becomes aware of the deficiency; information as to the delay suffered and the resulting costs can then be given as the extent of the delay, and its likely effects on the remainder of the work, become known.

When all information needed by the engineer to ascertain the validity and the evaluation (or 'quantum') of the claim has been submitted, the contractor is entitled to expect that it will be dealt with, so that any sum to which the contractor is found to be entitled may be included in a certificate as soon as may be practicable. It is to be expected that the engineer may require a reasonable period of time for his consideration of the basis and evaluation of the claim; the question of what is reasonable depending upon the circumstances and the complexity of the claim. In most cases it should be possible for him to reach a decision within a month or two of his receiving full details, and where, because of the continuing nature of the additional costs or for any other reason, it is not possible to arrive at a final valuation the engineer should be willing to certify an appropriate payment on account. Certainly there is no excuse for any suggestion that claims should be considered after all of the work has been completed: indeed any question as to the validity of the claim should be raised immediately, so that it may be dealt with while the facts are fresh in the minds of those involved, and while, where necessary, subsoil or other information having a bearing upon the claim may be ascertained and recorded.

One of the difficulties facing the engineer in considering the contractor's claims is that he is called upon to play two quite different, and sometimes conflicting, roles. He must represent the interests of the employer, his client, of whom he is a paid agent; simultaneously he must act in a quasi-judicial capacity in maintaining a fair and impartial balance between the interests of the employer and of the contractor. In some cases that difficulty may be enhanced by the knowledge that his own actions as engineer under the contract may have given rise to the claim as, for example, where the contractor alleges that delay has been caused through late issue of working drawings.

During the course of the engineer's consideration of a claim the contractor should of course provide any additional information that the engineer may reasonably require, and he should recognize that in the event of his having to invoke arbitration, full details of the contractor's pricing or other relevant information will have to be produced on discovery. There is nothing to gain by any claim that the contractor's pricing details are 'confidential', although he may reasonably expect the engineer to preserve that confidentiality so far as may be consistent with his having access to all relevant information.

DISPUTES

If the engineeer fails to deal with the claims within a reasonable time, or rejects them in whole or in part, the contractor may (i) accept that action or inaction; (ii) seek an opportunity to persuade the engineer to reconsider the claims or to reverse his decision; (iii) give notice that a dispute has arisen.

Usually it is unwise for the contractor to take the third course until he is sure that there is nothing to gain by further discussion for although invoking the dispute procedure under Clause 66 of the ICE Conditions, or corresponding clause of another form, does not preclude further discussion, it may inhibit further negotiations towards the settlement of the claim by agreement. Employers often consider that receipt of a notice under Clause 66 is the moment at which the dispute should be put into the hands of their lawyers: thereafter negotiations are often conducted through the parties' lawyers, who may not have been given the authority to settle at a reasonable evaluation of the claims. Furthermore, the introduction of additional links in the chain of communication between employer and contractor tends to discourage constructive negotiation.

Where, after due consideration, the contractor decides to invoke Clause 66 he should write to the engineer requiring him to give his decision under that clause on the matters in difference, sending his letter by recorded delivery as a precaution against any possible future allegation that the notice was not received. At the same time the contractor should note in his diary the day on which three months* will have elapsed from the date of service of the notice.

During this first period of three months* the contractor should continue to provide any further information the engineer may require, and should attend for discussions of the claim if called upon to do so. After the period has elapsed, or upon receipt of the engineer's decision under Clause 66 if this arrives earlier, the contractor may, if dissatisfied with that decision, give notice of arbitration. Such notice must be given within three months of the engineer's Clause 66 decision or of the latest date on which he could have given such decision: or, where the ICE conciliation procedure has been invoked, within one month of the date of receipt of the conciliator's recommendation.

Earlier editions of the ICE Conditions sought to postpone any reference to arbitration until after completion – or alleged completion – of the works. That constraint has, under the sixth edition, been removed (see p. 14).

Before embarking upon an 'immediate' arbitration, however, the contractor should consider whether or not this is in his best interests. The purpose of the Clause 66 limitation of immediate arbitration, albeit that it was largely ineffective, was to ensure that as far as is possible all matters in dispute are dealt with in a single arbitration, which must of necessity be deferred until work has been completed. In this way costs should be kept to a minimum. Conversely, if the contractor seeks to refer each dispute to arbitration as it arises, the total of costs incurred in the series of arbitrations that might result is likely to be substantially greater than those of a single reference covering all of the disputed items. In his award of costs the arbitrator should have regard not only to the outcome of the claims: he

* One month in certain cases: see p. 13 *et seq*.

should also consider whether or not the costs were necessary. Hence, unless satisfied that the contractor had no option but to refer each matter separately – for example, for reasons of cash flow – he would in such circumstances award costs, or a major part of them, against the contractor, even where the claims succeed.

SELECTION OF THE ARBITRATOR

As has been explained in Chapter 4, the contractor may sometimes be able to influence the choice of arbitrator, in that he may put forward one or more nominees for the employer's agreement: he may, where the employer makes counter-nominations, either agree to one of the nominees or reject them all; and he may, where it becomes necessary to apply for an appointment by the President of the ICE or other appointing authority, suggest appropriate qualifications for the arbitrator.

It is, of course, fundamental that the arbitrator must be utterly impartial, and in putting forward nominees the contractor should observe this requirement strictly. The other basic requirements of the arbitrator are that he must have a general technical knowledge of the subject-matter of the dispute, but no knowledge of the particular dispute, and he must have judicial capacity.

Additionally, it is desirable from both parties' viewpoints that he should be knowledgeable and experienced in arbitration procedure and law, and, from the viewpoint of the contractor in particular, that he should have knowledge and understanding of the business of contracting.

Much detailed information as to these attributes is available in the List of Arbitrators, published by the Institution of Civil Engineers and available from the Arbitration Officer of that Institution for a small charge.

In assessing a candidate's knowledge of arbitration account should be taken not only of the statement of arbitration experience included in the list: the question whether or not the person is a Fellow of the Chartered Institute of Arbitrators is also relevant. For it is reasonable to assume that a person who is a Fellow of the CIArb recognizes arbitration as a subject worthy of study as a separate discipline, and that he has attained a standard of knowledge recognized by the Institute as being appropriate to one seeking appointment to be an arbitrator. Conversely, the appointment of an arbitrator who is not adequately trained in arbitration is likely to lead to additional costs and delays, while that person seeks advice on procedure or on legal questions that may arise during the reference, the answer to which would be known by an experienced arbitrator.

Where attempts by the parties to agree upon an arbitrator prove unsuccessful the contractor should ensure, when applying for an appointment by the President, that full information as to the matters in dispute is given on the application form, to assist him in making an appropriate choice. Where, for example, the dispute relates to the pricing of varied work, it is to be hoped that the arbitrator appointed will have experience in such matters: preferably as a contractor.

THE PRELIMINARIES

The main objectives of the claimant – in this case the contractor – are to obtain full recompense for the losses he has suffered, including the costs he incurs in obtaining that recompense, and to do so within as short a period of time as may be practicable. The first of these objectives is achieved by thorough preparation and presentation of the claimant's case, and the second may to a large extent be influenced by the claimant's action in setting the pace.

If, as sometimes happens, no communication is received from the arbitrator for a week or two after his appointment, the claimant should take the initiative by writing to him requesting him to convene a preliminary meeting as soon as possible, and putting forward up to say ten dates upon which the claimant would be able to attend. A copy of this letter, and indeed of all letters sent to the arbitrator, should be sent to the respondent, the original being marked to show that the copy has been sent. If the claimant is able he may offer to make available a suitable venue for the preliminary meeting, having regard to the convenience of the arbitrator and of the respondent. The claimant should, however, make it clear that the offer is made only in an attempt to be helpful, and that he is quite willing to attend at whatever venue may be chosen by the arbitrator.

If for any reason it appears inappropriate to suggest a meeting – for example where the sum in dispute is so small as not to warrant the costs involved – the contractor may write to the respondent suggesting a draft Order for Directions (see SD/16) for his agreement. If the respondent does agree, then the claimant may send a copy of the draft, and of the respondent's letter of agreement, to the arbitrator, requesting him to make the order by consent.

Where a preliminary meeting is held one of its main purposes is to define a timetable for the several stages of the interlocutory proceedings. The claimant should estimate in advance of the meeting how long he requires for preparation of his Points of Claim, bearing in mind that whatever time he requires will to some extent form a yardstick by which the respondent will request and the arbitrator will allow time for the Points of Defence. The respondent will probably argue in any case for a longer period of time than that allowed for the claim, on the ground that the claimant has been able to commence his drafting of the Points of Claim whenever he chose to do so, while the respondent must await receipt of that document before he can start to draft his defence. However in the case of disputes referred to arbitration under Clause 66 of the ICE Conditions it may usually be argued by the claimant that the respondent is already aware of the substance of the claims, which must already have been submitted to the engineer under the first stage of the Clause 66 procedure.

If the respondent requests what appears to the claimant to be an excessive period of time for preparing the defence he may request that the arbitrator reduces it to a reasonable period: but pressing such an argument too far may be inadvisable, because it may lead to requests at a later date for extensions of time.

In most cases where there is a counterclaim the claimant will be aware of its existence and substance before the preliminary meeting, and should therefore be able to estimate what period of time he will require for preparing his defence to it.

Points of Reply must be confined to dealing with any fresh matter raised in the defence, and therefore should not take long to prepare, if indeed they are needed at all. In many cases a period of two weeks is ample.

Time can often be saved by the parties' preparing their lists of documents concurrently with the final stages of the pleadings. Usually by that time it has become clear which documents are relevant to the matters at issue, and it is therefore not unreasonable to suggest that lists of documents are exchanged not later than the date fixed for the delivery of the Points of Reply.

Progress with the interlocutory proceedings is often delayed by requests for extensions of time. The claimant should, wherever possible, avoid having to ask for extensions, not only because of the delay they may, if granted, cause directly, but because of their effect in implying an entitlement of the respondent to a similar or perhaps longer extension. Usually the arbitrator will grant a first request, if reasonable in amount and soundly based, but in the case of subsequent applications by the same party he should generally seek the observations of the opposing party before determining the application, and should where appropriate do so at a meeting convened for that purpose. It often happens that the respondent, having no incentive to expedite the proceedings, seeks to delay them in every way open to him; and where the claimant becomes aware of that strategy he should if necessary draw it to the arbitrator's attention. The respondent in that situation will of course readily agree to any application by the claimant for an extension of time and will quote the extension given as a reason why he himself should be granted a similar or longer extension. Where the claimant wishes to oppose an application by the respondent for an extension, and where he can reasonably do so, he may request that the arbitrator hear his objections to the application at a meeting: and he may request a meeting in a case where the respondent has overrun the time allowed him for delivery of pleadings or other documents. In either of these situations the claimant may reasonably expect, and should request, that the arbitrator awards costs in his favour in any event: that is, whatever the outcome of the claims.

Another potential source of delay during the interlocutory proceedings is the possibility of requests for Further and Better Particulars. These can best be avoided by the claimant by ensuring that his Points of Claim are fully detailed, and that they do not include any vague allegations: for example that the engineer ordered 'numerous' variations, or that he failed on 'several' occasions to reply to the claimant's request for information. Such expressions invite a request for details of all of the variations alleged to have been ordered, and of each and every occasion upon which the contractor requested, but did not receive, information. By foreseeing such requests for particulars, and by including them in the original pleadings, the scope for delaying tactics may be restricted.

On the other hand the contractor must ensure that he has all the particulars that may be needed in refuting any allegations in the defence. Where, for example, there is an allegation that the works were delayed, not by a failure by the engineer to provide working drawings but by insufficiency of plant and labour on the site, the contractor should ensure that he has full records of the plant used, with dates of arrival on and departure from the site of each item, and the numbers of each category of labour on site during each week of the contract period. Where such matters are likely to be relevant, the claimant should summarize the information and request that the respondent agree it, having examined the original records from which it is extracted. In this way unnecessary and expensive delays during the hearing, while details are checked, may sometimes be avoided.

Another important part of the preliminaries is that of discovery, in that it provides an opportunity to ascertain the truth. The contractor should ensure that all documents likely to be relevant are produced, subject only to the limitation of privilege where relevant. For example, where a subcontractor is in dispute with the main contractor, correspondence between the main contractor and the engineer or the employer may be very relevant to any allegation of defects in the subcontractor's work, and to any allegation by the subcontractor that his claims have not been adequately presented to the employer.

THE HEARING

In his preparations for the hearing the contractor should study the pleadings carefully to determine the precise points at which issues arise, and should prepare his evidence to prove those issues. Much valuable time is saved at the hearing by omitting any matter that is not in contention, except as may be needed to form a background to the matters in dispute. He can then concentrate his attention on the matters he has to prove, and ensure that adequate evidence is available on each such matter.

Documentary evidence carries more weight than evidence given orally, because it is not subject to the vagaries of memory or to deliberate distortion. Where credibility is likely to be in issue, the contractor should if necessary request that the arbitrator takes evidence on oath – which he should generally do without prompting from a party – and he may where necessary go further, in deciding to be represented in such cases by counsel. This is because counsel's skill in cross-examination may be invaluable in ascertaining the truth where witnesses are suspected of being unreliable. Again, the appointment of counsel may be wise in cases where difficult questions of law arise. In many cases, however, the issues that arise are mainly of a technical nature, and as such may be better presented to the arbitrator by an expert or by an advocate who understands those technicalities, provided that that expert or advocate is also skilled as an expert witness or as a technical advocate, as the case may be.

REQUEST FOR REASONED AWARD

It is now standard practice for an arbitrator to include reasons in his award, and usually both parties are content that he should do so: not necessarily because they expect to mount a challenge to the arbitrator's decision on any controversial question of law that may have arisen, but simply so that the losing party knows why it has lost. However, there may be special reasons why a party may wish to safeguard, or to waive, its right to have a reasoned award.

Where some important question of law arises, and where a party is determined to challenge a possibly adverse decision on that question, it should expressly *require* of the arbitrator that his award includes reasons, whether or not the arbitrator indicates his intention to give such reasons. By so doing the party's rights under section 1(5) and (6) of the 1979 Act are protected (see p. 78).

Conversely, where the principal aim of a party is to bring the dispute to finality, win or lose, that party may have suggested to its opponent that they enter into an exclusion agreement (see p. 12). Alternatively, the party may suggest to its opponent that the arbitrator be requested *not* to give reasons. Such a request is, however, unlikely to be effective unless both parties so agree.

OFFERS

At any stage during the proceedings the respondent may make an offer to settle the claims. Besides his consideration of the acceptability of that offer, the contractor should consider the significance of it as a factor influencing the award of costs. He should form an assessment of the value of the claims in total, having regard to the probability of success in the case of each item, and the likely award against that item if it succeeds. Where the offer made is in excess of the evaluation of claims made in this way, and taking account of costs incurred up to the time of the offer, then the contractor should accept the offer: otherwise he incurs the very real risk that, should the arbitrator award a lower sum, he may order the contractor to bear both parties' costs and the costs of the award – that is, the arbitrator's charges – from the date of the offer.

Where the contractor prefers to gamble upon being awarded a higher sum by the arbitrator, or where he thinks it likely that the respondent may increase his offer, he may of course reject an offer that appears to be in his favour. Alternatively, where there are a number of items in the claim, he may request that the respondent makes a specific offer against each item, in the hope that agreement can be reached on some of the claims, leaving the others to be dealt with by the arbitrator.

UPSETTING THE AWARD

Where the claimant believes that the award is wrong in law, or that the arbitrator has misconducted himself or the proceedings he may, subject to the limitations referred to in Chapter 10, initiate proceedings for an appeal for remission or setting aside of the award. Any such action must be taken within 21 days of the date of publication of the award (RSC Order 73 Rule 5 as amended: see Appendix E).

Appendix A: Specimen documents

SD/1 Application for stay of court proceedings
SD/2 Request for engineer's decision under Clause 66 of ICE Conditions
SD/3 Form ArbICE (revised): Part 1: The Contract(s)
SD/4 Form ArbICE (revised): Part 2: Notice to Refer
SD/5 Form ArbICE (revised): Parts 3 & 4: Notice to Concur & Application to the President to appoint an Arbitrator
SD/6 Form ArbICE (revised): Part 5: Appointment by President
SD/7 Notice of Arbitration and Notice to Concur: general
SD/8 Application to President to appoint: general
SD/9 Notice of Appointment: general
SD/10 Notice of Appointment: Small Claim: Attended Hearing
SD/11 Notice of Appointment: Small Claim: Documents Only
SD/12 Notice of intention to proceed *ex parte*
SD/13 Arbitration Agreement
SD/14 Exclusion Agreement
SD/15 Agenda for Preliminary Meeting
SD/16 Order for Directions
SD/17 Points of Claim
SD/18 Request for Further and Better Particulars
SD/19 Further and Better Particulars
SD/20 Points of Defence and Counterclaim
SD/21 Points of Reply and Defence to Counterclaim
SD/22 Scott Schedule
SD/23 List of Documents
SD/24 Calderbank letter
SD/25 Proof of Evidence
SD/26 Interim Award
SD/27 Final Award
SD/28 Notice of publication of Award
SD/29 Letter of Intent

SD/1 Application for stay of court proceedings

IN THE HIGH COURT OF JUSTICE 1993 W No 1234
QUEEN'S BENCH DIVISION

BETWEEN:

Wright, Charlie & Company
PLAINTIFFS

and

The Universal Construction Company Limited
DEFENDANTS

1. This is an application by the Defendants in this action for a stay of the proceedings under Section 4 of the Arbitration Act 1950.

2. The matters in dispute arise from a Contract in writing between the parties dated the 25th day of June 1992.

3. The said Contract provides, in Clause 18 thereof, a provision that any dispute between the parties that may arise from the Contract shall be referred to the arbitration and final decision of a person to be agreed upon between the parties or, failing agreement, to be appointed upon the application of either party by the President for the time being of the Chartered Institute of Arbitrators.

4. The Defendant has entered an appearance to the writ in this action and has a good defence thereto.

5. At the date when the action was commenced the Defendant was, and the Defendant remains, ready and willing to do all things necessary for the proper conduct of the arbitration, in accordance with the provisions of the said Contract.

Dated this First day of July 1993

Joe Bloggs

For and on behalf of the Universal Construction Company Limited

Defendants in the action.

SD/2 Request for engineer's decision under Clause 66 of ICE Conditions

From: The Universal Construction Company Limited

To: I K Brunel Esq

2nd January 1996

RECORDED DELIVERY

Dear Sir

Construction of Tunnel under the English Channel

We hereby give notice that a dispute has arisen from the above contract as a consequence of your failure to certify payment of sums due to us in respect of claims totalling £56m in value, full details of which were notified to you in our submission reference UCC/Chunnel/Claim under cover of our letter dated 1st July 1995.

We request your decision, under Clause 66 of the Contract, as to whether or not you will certify payment to us of the said sum.

Yours faithfully

For the Universal Construction Company Limited

J Bloggs
Managing Director

SD/3 Form ArbICE (revised): Part 1: The Contract(s)

Form ArbICE (revised) issued by **The Institution of Civil Engineers**
Great George Street **London SW1P 3AA**
Telephone: 01-222-7722 Telegrams: Institution London SW1

These Forms are intended for use with the *ICE Arbitration Procedure*[1] in connection with arbitrations under the *ICE Conditions of Contract* or the *FCEC Form of Sub-Contract*. They are, however, easily adapted for use in other arbitrations.

Part 1: THE CONTRACT(S)

Title of (Main) Contract: Tunnel under the English Channel

Date of (Main) Contract: ~~10th December 1992~~

Brief description of the Works:
Construction of concrete-lined tunnel between Dover and Calais

Employer's name: Department of Transport

Employer's address: Marsham Street, London SW1

(Main) Contractor's name: The Universal Construction Company Limited

(Main) Contractor's address: Telford House Basingshot Hampshire

Engineer's name: I K Brunel

Engineer's address: The Manse, Clifton, Bristol

Title of Sub-Contract (if any)[2]:

Date of Sub-Contract[2]:

Brief description of the Sub-Contract Works[2]:

Sub-Contractor's name[2]:

Sub-Contractor's address[2]:

Please state the form of the Arbitration Clause:

Clause 66 of the *ICE Conditions of Contract* ✔

~~Clause 18 of the *FCEC Form of Sub-Contract*~~

~~Clause 67 of the *FIDIC Conditions of Contract*~~

~~Other (please state, and attach a copy hereto).~~

[1] The *Institution of Civil Engineers' Arbitration Procedure 1983* (England & Wales) or (Scotland) may be obtained from the marketing department at the Institution, price £1.50. Cheques should be made payable to the Institution of Civil Engineers.
[2] Delete as appropriate.

SD/4 Form ArbICE (revised) : Part 2 : Notice to Refer

Part 2: NOTICE TO REFER[1] DISPUTE(S) OR DIFFERENCE(S) TO ARBITRATION.

WHEREAS dispute(s) or difference(s) as hereinafter described have arisen between the Parties to the Main Contract[2]/~~Sub-Contract~~[2] (as described in Part 1 hereto) in connection with or arising out of the said Contract or the carrying out of the Works

NOW we the undersigned hereby give notice requiring the said dispute(s) or difference(s) to be referred to arbitration.

Dated this Twentieth day of April 19 96

Signed [signature]

for and on behalf of The Universal Construction Company Limited

the ~~Employer~~[2]/(Main) Contractor[2]/~~Sub-Contractor~~[2].

Brief description of the dispute(s) or difference(s) to be referred to arbitration:

Rejection of Main Contractor's claim for additional costs and delay incurred in dealing with unforeseeable physical conditions and artificial obstructions

To be completed where Clause 66 of the *ICE Conditions of Contract* or Clause 67 of the *FIDIC Conditions of Contract* apply.

The above dispute(s) or difference(s) were referred to the Engineer in accordance with Clause 66/~~67~~[2] of the Conditions of Contract in letter(s)

from The Universal Construction Company Limited

dated 2nd January 1996

Notice of the Engineer's decision thereon was given in the Engineer's letter(s)[2]

to The Universal Construction Company Limited

dated 26th March 1996

~~The Engineer has so far failed to give notice of his decision thereon.~~[2]

[1] The *Institution of Civil Engineers' Arbitration Procedure 1983* (England & Wales) or (Scotland) may be obtained from the marketing department at the Institution, price £1.50. Cheques should be made payable to the Institution of Civil Engineers.

[2] Delete as appropriate.

SD/5 Form ArbICE (revised) : Parts 3 & 4 : Notice to Concur & Application to the President to appoint an Arbitrator

Part 3: NOTICE TO CONCUR[1] IN THE APPOINTMENT OF AN ARBITRATOR.[2]

To: Department of Transport

Address: Marsham Street, London

WE HEREBY call upon you to concur in the appointment of an Arbitrator to hear and determine the dispute(s) or difference(s) between us as set out in Part 2 hereof.

WE PROPOSE the following person[3] for your consideration and require you within ~~14~~ 28 days of the service of this Notice
(i) to agree in writing to his appointment; or
(ii) to propose an alternative person for our consideration failing which we intend to apply to the President of the Institution of Civil Engineers to appoint an Arbitrator.

Name of person proposed as Arbitrator: Thomas Telford Esq FICE FCIArb

Address of person proposed as Arbitrator: Menai Bridge Road, Bangor

Dated this Twentieth day of April 1996

for and on behalf of The Universal Construction Company Limited

the ~~Employer~~[4]/(Main) Contractor[4]/~~Sub Contractor~~[4].

Signed... Joe Bloggs

Part 4: APPLICATION[5] TO THE PRESIDENT OF THE INSTITUTION OF CIVIL ENGINEERS TO APPOINT AN ARBITRATOR.[2]

To:
The President, the Institution of Civil Engineers
The Institution of Civil Engineers
Great George Street
London
SW1P 3AA

IN THE MATTER OF THE DISPUTE(S) OR DIFFERENCE(S) referred to in Part 2 of this Form and since the Parties have failed to agree upon an Arbitrator we hereby apply to you to appoint an Arbitrator.

We enclose a cheque[3] for £ 80 plus VAT in respect of the charge made by the Institution towards administrative costs in connection with this application.

We think it desirable that the Arbitrator should if possible ~~be skilled~~/have experience[4] in the following fields or professions: Tunnelling; engineering geology; evaluation of varied work

The amount at issue is approximately £ 56m

Dated this Twentyfifth day of May 1996

Signed Joe Bloggs

for and on behalf of The Universal Construction Company Limited

the ~~Employer~~[4]/(Main) Contractor[4]/~~Sub Contractor~~[4].

[1] See Rule 2 of the *ICE Arbitration Procedure*.
[2] In a Scottish arbitration the word 'Arbitrator' shall mean 'Arbiter'.
[3] The Institution publishes a *List of Arbitrators* which includes information on the arbitration experience and careers of the persons listed. The list may be used by parties seeking to reach agreement on the choice of an arbitrator and is obtainable from the arbitration office at the Institution, price £1.50. Cheques should be made payable to the Institution of Civil Engineers.
[4] Delete as appropriate.
[5] See Rule 3 of the *ICE Arbitration Procedure*.

SD/6 Form ArbICE (revised) : Part 5 : Appointment by President

Part 5: APPOINTMENT OF AN ARBITRATOR[1] BY THE PRESIDENT OF THE INSTITUTION OF CIVIL ENGINEERS.

From the President: Sir John Rennie FRS FICE

To: The Universal Construction Company Limited

Copies for information to: The Department of Transport

and: Mr Stephenson

I hereby appoint: Robert Stephenson FRS FICE FCIArb

of: 1 High Street Newcastle-on-Tyne

Arbitrator in this matter and I hereby direct that the Arbitration be conducted in accordance with the *ICE Arbitration Procedure 1983* (England & Wales) ~~or (Scotland).~~

Dated this Thirtieth day of June 1996

[signature]

PRESIDENT/~~VICE PRESIDENT~~

[1] In a Scottish arbitration the word 'Arbitrator' shall mean 'Arbiter'.

SD/7 Notice of Arbitration and Notice to Concur : general

From: The Speedybuild Construction Company Limited

To: The Gigantic Property Development Company Limited

15th March 1993

Dear Sirs

Construction of Tower Block, Fred Needle Street

We hereby give notice that a dispute has arisen from our contract with you for the construction of the above building, by reason of the failure of your architect to certify payment of certain sums, details of which were submitted to him under cover of our letter of 1st December 1992.

We require that the dispute be referred to arbitration in accordance with the provisions of Clause X of the Contract.

In accordance with that clause we submit the names of the following gentlemen and we request that you concur in the appointment of one of them to be arbitrator in this reference:

Mr C Wren FRIBA FCIArb of St Paul's Chambers, London, or

Mr I Measure FRICS FCIArb of Yardstick Road, Finchley, or

Mr Archibald Tect RIBA FCIArb of High Street, Lowtown.

In the event that we do not receive a reply to this letter or that we are unable to reach agreement with you as to the appointment of a suitable person within 30 days of the date of this letter we shall apply in accordance with the said Clause X to the President of the Chartered Institute of Arbitrators for an appointment to be made by him.

Yours faithfully

for the Speedybuild Construction Company Limited

P O'Reilly

Director

SD/8 Application to President to appoint: General

From: The Speedybuild Construction Company Limited

To: The President: the Chartered Institute of Arbitrators

Copy to: The Gigantic Property Development Company Limited

20th April 1993

Dear Sir

Construction of Tower Block, Fred Needle Street

We are under contract with the Gigantic Property Development Company Limited to construct the above building, work on which is in our opinion substantially completed. The contract is on the XYZ standard form which as you will know provides for the appointment of an arbitrator by yourself in the event that the parties are unable to agree upon such appointment.

A dispute having arisen we wrote to the Employer on 15th March 1993 giving notice of arbitration and notice to concur, but we have received no reply. A copy of our letter of 15th March 1993 is enclosed.

Accordingly we request that you appoint an arbitrator to determine the matters in dispute. These matters concern the evaluation of additional excavation and concrete in deep foundations, including the construction of a cofferdam, heavy reinforced concrete foundations and superstructure, and the evaluation of additional costs incurred as a result of delays in respect of late variations and additional works.

The sum in dispute is approximately £1.7m plus interest. We suggest that the arbitrator should be either an architect or a quantity surveyor with experience of large office block construction in London.

Yours faithfully

for the Speedybuild Construction Company Limited

P O'Reilly

Director

SD/9 Notice of Appointment: General

From: Robert Stephenson FRS FICE FCIArb 10 July 1996

To: The Universal Construction Company Limited

The Department of Transport

Gentlemen:

Arbitration Between the Universal Construction Company Limited *(Claimant)* - and - The Department of Transport *(Respondent)*: Tunnel under the English Channel

I am appointed by the President of the Institution of Civil Engineers to be Arbitrator in the above reference. I hereby accept the appointment.

Proposed terms of my appointment and basis of my charges are set out in the Form of Agreement appended hereto, which I have based on a standard form issued by the Society of Construction Arbitrators. Both parties are requested to sign the form and to return it to me.

In the event that either or both parties fail or refuse to return the form, duly signed, to me within twentyeight days of the date of this letter I shall in due course exercise my power under Section 18 (1) of the Arbitration Act 1950 to tax the costs of the award: and meanwhile I shall require security for those costs.

It is my intention to convene a Preliminary Meeting for Directions during the week commencing on 30 July 1996 and I require that both parties advise me no later than 17 July 1996 of any dates during that week that they would find unsuitable for a meeting commencing at 11.00 am. I propose that such meeting be held at the Institution of Civil Engineers, 1 Great George Street, Westminster SW1P 3AA, unless both parties agree upon some other venue and advise me as to their wishes. I am willing to travel to whatever venue may be agreed between the parties as being the most convenient to them.

When replying to this letter, either by post or by fax, and when writing to me in the future, please send a copy of your letter to the other party, indicating on your letter to me that such copy has been sent. Please do not attempt to communicate with me by telephone.

Yours faithfully

Arbitrator

THE SOCIETY OF CONSTRUCTION ARBITRATORS

FORM OF AGREEMENT

(for use where an Arbitrator
is appointed by
an Appointing Authority)

By the terms of an Agreement in writing dated the day of 19

between of

and of

[and of]

relating to

it is provided that any dispute or difference arising thereunder shall be referred to arbitration.

A dispute or difference having arisen between the parties to the said Agreement, and

of

(hereinafter referred to as "the arbitrator") having been appointed by **

to be sole arbitrator in this matter pursuant to the provisions of

the said Agreement, which appointment the arbitrator [*has accepted] [*will accept upon agreement of the terms herein set out], it is hereby agreed that the following Terms and Conditions shall apply.

** Here insert the name of the Appointing Authority.

* Delete as appropriate.

Fees

1. Notwithstanding any order as to costs in any award, the parties shall be jointly and severally liable to the arbitrator for the due and timely payment of his fees, costs and expenses in accordance with these Terms and Conditions.

2. Whether or not a hearing is conducted or an award made, the arbitrator shall be entitled to payments at the rates set out herein for all time spent on or connection with the arbitration and for all time allocated thereto, together with all out-of-pocket expenses and other disbursements reasonably incurred. Where applicable, value added tax shall be charged and paid for in addition to such fees, expenses and disbursements.

3. The arbitrator's fees shall include the following:

(a) Either (i) an **appointment fee** of £

or (ii) a **minimum fee** of £

which shall be paid to and retained by the arbitrator in any event.

(b) An **hourly rate** of £ for each hour or part of an hour spent on or in connection with the arbitration, including but not limited to time spent on general administration, travelling and reading papers.

Provided that, if no daily rate is specified in paragraph 3(c)(i) below, all time spent on a day set down for hearing the arbitration (including time spent in adjournments) shall rank for payment at the hourly rate, with a minimum of eight hours' payment for that day in any event.

(c) (i) A **daily rate** of £ for each day or part of a day spent in hearings, meetings, inspections, site visits and the like. A day shall be deemed to comprise up to but no more than eight hours including adjournments. Time spent in excess of this limit shall be charged extra at the hourly rate, as shall time spent in travelling and reading papers on a hearing day.

(ii) Where appropriate, an **alternative daily rate** of £ for each day not occupied in arbitration business which is necessarily spent away from the arbitrator's normal residence or place of business. Where a day is charged for on this basis, any arbitration work actually undertaken on that day shall be deemed to be covered thereby.

(d) A **booking fee** of £ for each day reserved for a hearing, payable by the party or parties requesting the booking and when the booking is made. The arbitrator shall give credit for any booking fees he has received against any other fees which may subsequently become due in respect of the days so reserved.

(e) A **cancellation fee** for each day reserved for a hearing which is later vacated, whether by adjournment or by cancellation, according to the following scale:

(i) Where the booking is vacated more than six months before the first day so reserved:
A fee equal to per cent of the daily rate for each day vacated.

(ii) Where the booking is vacated six months or less but more than three months before the first day so reserved:
A fee equal to per cent of the daily rate for each day vacated.

(iii) Where the booking is vacated three months or less but more than one month before the first day so reserved:
A fee equal to per cent of the daily rate for each day vacated.

(iv) Where the booking is vacated one month or less before the first day so reserved:
A fee equal to per cent of the daily rate for each day vacated.

For the purpose of this paragraph the daily rate shall be that specified in paragraph 3(c)(i) above or, if none is there specified, a rate equivalent to eight times the hourly rate specified in paragraph 3(b) above.

The above scale shall apply whatever the length of the booking vacated. Provided that, if a booking is vacated due to circumstances over which the parties have no control, any booking fee already paid shall be refunded.

Payment

4. Notwithstanding any provision for a payment by way of security, the arbitrator shall be entitled to submit, and the parties shall pay, such interim accounts of fees, expenses and disbursements as the arbitrator thinks appropriate.

5. Where the arbitration continues for more than one year after the arbitrator was appointed, the arbitrator shall be entitled to re-value his fees in respect of subsequent years generally in line with the General Index of Retail Prices (RPI).

6. Payment in full of the arbitrator's fees, expenses and disbursements shall be made by the parties within ten days of the date of publication of each and every award, or within forty days of the date of submission of each interim account rendered in accordance with paragraph 4, above. Should the parties fail to make payment in full within the said ten or forty days (as the case may be), interest upon all monies then outstanding shall accrue at a rate of per cent above the base rate charged from time to time by the arbitrator's Bankers.

General

7. The arbitrator shall be entitled to require any or all of the parties to pay such amount as he shall determine from time to time to be held by the arbitrator as security for the due payment of his fees, expenses and disbursements.

8. The arbitrator's disbursements may include the cost of obtaining such legal advice or technical assistance as in his absolute discretion he shall deem it desirable to take.
Provided that the advice of leading counsel shall not be sought without the express agreement of all the parties to the arbitration.

9. Should a settlement be reached between all or any two of the parties, and whether such settlement disposes of all or only some of the matters at issue, the parties shall forthwith inform the arbitrator of the terms of the settlement so that the same may be incorporated in an award.

Signed and delivered by:

______________________	______________________	______________________
on behalf of	on behalf of	on behalf of
date	date	date

Signed and delivered by the said arbitrator

____

date

SD/10 Notice of Appointment: Small Claim: Attended Hearing

From: P R Oberty MA FICE FCIArb 20 July 1992

To: C L Aimant Esq
R E Spondent Esq

Gentlemen

Arbitration between Claude Leslie Amaint - and - Roger Ernest Spondent

I am appointed the President of the Chartered Institute of Arbitrators to be Arbitrator in the above reference. I hereby accept the appointment.

My charges will be at the rate of £X per hour for the time during which I engage myself upon, or which I allocate to, the duties of the reference, plus the amount of all expenses incurred in the execution of those duties, plus Value Added Tax on the amount of my fees and expenses.

In connexion with the appointment I have received a copy of the contract between the parties and of the Claimant's letter of application dated 12 May 1992 to the President of the Chartered Institute of Arbitrators.

I give below a brief explanation of procedure in arbitration, and indicate a time-table for preliminary matters which I propose, subject to my considering any application either party may wish to make, to adopt in this reference.

1. Representation

It is open to each party either to conduct its own case or to be represented professionally: for example by a technical person or by a lawyer. Each party must however give notice of its intention in order to give the other party an opportunity to arrange similar representation if that party so wishes.

In considering this point the parties should recognise that professional representation may be costly. In making my Award in due course it is a part of my duty to award costs, at my discretion; and in exercising that discretion I shall have regard both to the outcome of the claim and to the question whether or not costs were incurred necessarily.

2. Pleadings

English law requires that each party shall be forewarned of the case to be presented by the other party. This requirement is met by an exchange of Pleadings, which comprise Points of Claim, Points of Defence, and where necessary Points of Reply. If there is a Counterclaim it is pleaded with the Points of Defence, and the Defence to the Counterclaim is pleaded with the Points of Reply. Each document in the Pleadings must set out, in summary form, the material facts upon which the party pleading intends to rely, and must be sent to the other party and copied to me.

The Points of Claim must state in itemised form the nature of each claim, the sum claimed in respect of each item, and the amount of the claim; and must be submitted within 14 days of the date of this letter.

Within 14 days of receipt of the Points of Claim the Respondent must submit its Points of Defence, which should deal with each and every allegation in the Points of Claim, stating whether the allegation is admitted, not admitted or denied. Whether or not an item claimed is admitted in principle the Respondent may challenge the sum claimed in respect of that item. The Points of Defence may include other statements.

Thereafter the Claimant may, within a further 7 days, submit its Points of Reply, which must deal only with matters raised in the Points of Defence.

3. Discovery and Inspection

Both parties will be required in due course to disclose all documents relevant to the matters in question in this reference, which are or have been in its possession, whether or not the documents support the case of the party holding them.

Within 7 days of delivery of the last document in the Pleadings - usually the Points of Reply - each party must send to the other party and to me a list of all of its relevant documents. Such documents, with certain limitations relating to privilege which I shall if necessary explain, must be made available for inspection or for copying by the other party, and must be brought to the Hearing.

4. Hearing

Subject to the timely completion of the above preliminaries it should be possible to convene a Hearing during October 1992. I propose that such Hearing be held at the Chartered Institute of Arbitrators, 24 Angel Gate City Road London EC1V 2RS or such other suitable venue as may be agreed.

5. Other Matters

Both parties are required to acknowledge receipt of this letter and to state their intentions regarding representation, to which I refer in paragraph 1 above.

The parties should then proceed with the action referred to in paragraphs 2 and 3 above, after completion of which I shall give directions as to the date and time of the Hearing.

Should either party find difficulty in complying with the time limits given above they may apply to me for an extension, stating reasons for their application and the amount of additional time for which they apply. I will then consider the application and deal with it at my discretion.

I shall have no communication with either party without the knowledge of the other party, and I shall accordingly address all of my letters, which may be sent by post or by fax, to both parties.

When writing to me each party must send a copy of its letter to the other party, indicating to me that they have complied with this direction.

For similar reasons I shall have no communication by telephone with either party. Urgent communications shall where possible be sent by fax.

Yours faithfully

Arbitrator

SD/11 Notice of Appointment: Small Claim: Documents Only

From: P R Oberty MA FICE FCIArb 20 July 1992

To: C L Aimant Esq
R E Spondent Esq

Gentlemen

Arbitration between Claude Leslie Amaint - and - Roger Ernest Spondent

I am appointed the President of the Chartered Institute of Arbitrators to be Arbitrator in the above reference. I hereby accept the appointment.

My charges will be at the rate of £X per hour for the time during which I engage myself upon, or which I allocate to, the duties of the reference, plus the amount of all expenses incurred in the execution of those duties, plus Value Added Tax on the amount of my fees and expenses.

In connexion with the appointment I have received a copy of the contract between the parties and of the Claimant's letter dated 12 May 1992 to the President of the Chartered Institute of Arbitrators, applying for the appointment of an arbitrator.

Having regard to the small sum in dispute in this reference I suggest that the parties may wish to adopt, by agreement, the following rules of procedure with a view to minimising costs:

1. That neither party be represented by a lawyer or an expert:

2. That the matters in dispute be determined upon my consideration of written submissions and evidence by the parties, and if necessary an inspection of the subject-matter of the dispute, in the presence of both parties. At any such inspection I would if necessary ask questions of either or both parties in order to clarify matters contained in the parties' written submissions:

3. That the parties comply with the following programme, subject to the right of either party to apply to me for an extension of time upon reasonable grounds and stating the extension requested:

   3.1 Within 14 days of the date by which both parties have agreed to adopt this procedure the Claimant shall submit to me, with a copy to the Respondent, a Statement of its Claim together with documentary evidence in support of that Claim and any other submissions that the Claimant may wish to make:

   3.2 Within 14 days of receipt of the above documents the Respondent shall submit to me, with a copy to the Claimant, a Statement of its Defence together with documentary evidence in support of that Defence and any other submissions that the Respondent may wish to make:

   3.3 Within 7 days thereafter the Claimant may submit to me with a copy to the Respondent a Reply to any matter raised in the Defence:

3.4 Thereafter on a date to be arranged I shall if necessary convene a meeting with both parties for the purpose of inspecting the subject-matter of the dispute. At any such meeting the parties would be permitted to draw to my attention any factual matters they may wish me to note, but not to make any further representation.

Both parties are required to advise me in writing within 14 days of receipt of this letter whether or not they agree to adopt the procedure set out in paragraphs 1 to 3 above. If and when both parties do so agree the Claimants should proceed in accordance with paragraph 3.1 above.

If either or both parties do not so agree it will become necessary to adopt more elaborate procedures about which I shall give directions should the need arise.

I do not intend to have any communication with either party without the knowledge of the other party. Accordingly I shall address all of my letters to both parties.

When writing to me, the parties must send a copy of their letter to the other party and must indicate on their letter to me that they have complied with this direction.

For similar reasons I shall not accept any telephone call from either party, and I shall not meet either party except in the presence of the other party.

Urgent communications may be sent to me by fax, provided that a copy of the communication is sent to the other party, if possible also by fax.

Yours faithfully

Arbitrator

SD/12 Notice of intention to proceed *ex parte*

From: P R Oberty MA FICE FCIArb 20 November 1992

To: C L Aimant Esq
R E Spondent Esq

RECORDED DELIVERY

PEREMPTORY

Gentlemen

Arbitration between Claude Leslie Aimant (*Claimant*) - and - Roger Ernest Spondent (*Respondent*)

In my letter of 20th July 1992 addressed to both parties I explained the basis of arbitration procedure and I gave directions in respect of this reference: including a direction that the parties acknowledge receipt of that letter and indicate their intentions as to representation.

The Claimant complied with my directions, but the Respondent did not.

In my letter of 20th August 1992 I advised the Respondent that if he wished to defend the claim against him he would have to comply with my directions and to submit his defence. He has failed to do so.

The Post Office has confirmed that my letter of 20th August and all of my subsequent letters were duly delivered to the Respondent's address.

Accordingly I now give this ***PEREMPTORY NOTICE*** of my intention to hear the Claim at Courtroom No 1, the Chartered Institute of Arbitrators, 24 Angel Gate, London EC1V 2RS on 15th December 1992, commencing at 10.30 am. In the event that the Respondent fails to appear or to be represented at the said hearing I shall, on the Claimant's application, proceed *ex parte* to hear the Claim and thereafter to make my Award.

Yours faithfully

Arbitrator

SD/13 Arbitration Agreement

IN THE MATTER OF THE ARBITRATION ACTS 1950–1979

AND

IN THE MATTER OF AN ARBITRATION BETWEEN

Majorspan Bridgebuilders PLC

CLAIMANT

AND

Universal Finance Corporation PLC

RESPONDENT

ARBITRATION AGREEMENT

We the undersigned HEREBY AGREE to refer to arbitration the dispute that has arisen from the contract between us constituted by the Claimant's tender dated 16 July 1993 offering for a consideration to carry out certain works of civil engineering construction and the Respondent's letter dated 30 July 1993 accepting that tender; and we HEREBY AGREE that the Arbitrator in the reference shall be Isambard Kingdom Brunel, Chartered Civil Engineer, of The Manse, Clifton, Bristol.

Signed by/on behalf of the Claimant by:

Dated this day of 1993

Signed by/on behalf of the Respondent by:

Dated this day of 1993

SD/14 Exclusion Agreement

IN THE MATTER OF THE ARBITRATION ACTS 1950–1979

AND

IN THE MATTER OF AN ARBITRATION BETWEEN

Majorspan Bridgebuilders PLC

CLAIMANT

AND

Universal Finance Corporation PLC

RESPONDENT

EXCLUSION AGREEMENT

We the undersigned having referred to arbitration a dispute that has arisen from the Contract between us constituted by the Claimant's tender dated 16 July 1993 and the Respondent's acceptance dated 30 July 1993 and having appointed Isambard Kingdom Brunel to be Arbitrator in the reference, HEREBY AGREE pursuant to Section 3 of the Arbitration Act 1979 that the jurisdiction of the High Court under Section 1 of the said Act with respect to any question of law arising out of any Award made in this reference and under Section 2 of the said Act with respect to any question of law arising in the course of the reference be excluded.

Signed by/on behalf of the Claimant by:

Dated this day of 1993

Signed by/on behalf of the Respondent by:

Dated this day of 1993

SD/15 Agenda for Preliminary Meeting

Arbitration between Majorspan Bridgebuilders PLC ***(Claimant)***

-and - Universal Finance Corporation PLC ***(Respondent)***

AGENDA FOR PRELIMINARY MEETING ON 20 August 1993

1. **APPEARANCES:**

	CLAIMANT	RESPONDENT
Name		
Appointment		
Name		
Appointment		

2. **CONTRACT:**

Form, edition etc

Date executed

Arbitration Clause and Procedural Law

Check validity of appointment

Copy for Arbitrator's records

3. **OUTLINE OF DISPUTE:**

Claim: subject

Approximate value

Counterclaim (if any): subject

:approximate value

	CLAIMANT	RESPONDENT
4. **FORM OF REPRESENTATION:**		

5. **RULES OF PROCEDURE:**

Exclusion Agreement?

Adoption of rules: CIArb/ICE/other?

ICE Short Procedure (omitting 21.1)?

Documents only?

Determination of issues in stages?

Request for reasoned award?

6. **PLEADINGS: TIME ALLOCATED:**

Points of Claim within	weeks
Points of Defence (and Counterclaim?) within	weeks thereafter
Points of Reply (and Defence to CC?) within	weeks thereafter
Reply to Defence to CC within	weeks thereafter

Requests for Particulars within 2 weeks of Pleading referred to

Particulars, where needed, to be provided within 2 weeks

Pleadings closed 1 week after service of last Pleading

7. **DISCOVERY: TIME ALLOCATED:**

List of Documents within	weeks of close of pleadings
Inspection within	weeks thereafter
Preparation of agreed bundles within	weeks thereafter.

8. **EXPERTS (IF ANY):**

Limitation as to numbers

Disciplines

Date for exchange of Reports

Meeting of experts of like disciplines

Date for submission of agreed Report and individual Reports

9. **PROOFS OF EVIDENCE: WITNESSES OF FACT:**

Date for exchange of proofs

10. **OPENING ADDRESS(ES) IN WRITING:**

Dates for submission

11. **HEARING:**

Commencement date

Number of working days to be reserved

Dates

Venue To be reserved by

Transcript or tape recording of evidence and submissions

Inspection of Real Evidence?

Closing addresses in writing?

Brief hearing, if necessary, after submission of closing addresses

12. **COSTS:**

Propose Interim Award on substantive matters and thereafter to provide opportunity for parties to address me on costs

13. **GENERAL DIRECTIONS:**

Copies of letters to Arbitrator, by post or by fax, to be sent to other party and originals to be marked accordingly

No communication with Arbitrator by telephone

Use fax, with copy to other party, for urgent communications

Figures to be agreed as figures where possible

Plans, photographs and documents to be agreed where possible

Liberty to apply

Costs of this application and order to be costs of reference.

14. **TERMS OF APPOINTMENT:**

Parties' agreement to proposed terms:

Claimant:

Respondent:

SD/16 Order for Directions

IN THE MATTER OF THE ARBITRATION ACTS 1950-1979

AND

IN THE MATTER OF AN ARBITRATION BETWEEN

Majorspan Bridgebuilders PLC

CLAIMANT

AND

Universal Finance Corporation PLC

RESPONDENT

ORDER FOR DIRECTIONS

Upon hearing the parties' representatives on both sides the following Directions are given and I hereby Order that:

1. There shall be Pleadings in this arbitration as follows:
   - 1.1 Points of Claim shall be served no later than 20 October 1993.
   - 1.2 Points of Defence and Counterclaim shall be served within 8 weeks thereafter.
   - 1.3 Points of Reply and Defence to Counterclaim shall be served within 6 weeks thereafter.
   - 1.4 Points of Reply to Defence to Counterclaim, if needed, shall be served within 2 weeks thereafter.
   - 1.5 Any request for Further and Better Particulars shall be served within 2 weeks of delivery of the Pleading referred to.
   - 1.6 Further and Better Particulars, where required, shall be served within 2 weeks of the request therefor.
   - 1.7 Pleadings shall be deemed to be closed 1 week after service of the last document therein.

2. There shall be Discovery of Documents as follows:
   - 2.1 Within 2 weeks of the close of Pleadings the Claimant and the Respondent shall each deliver to the other a List of Documents which are or have been in their possession or power relating to the matters in question in this arbitration:
   - 2.2 Inspection shall be given within 2 weeks thereafter:
   - 2.3 Agreed bundles of documents shall be prepared within 4 weeks thereafter.

3. Experts may be appointed by the parties provided that:
   - 3.1 No more than 2 Experts shall be appointed by either party, of whom one shall be a geologist and the other a quantity surveyor
   - 3.2 Reports of Experts of like disciplines shall be exchanged no later than 30 March 1994.
   - 3.3 Following upon such exchange Experts of like disciplines shall meet and shall prepare and submit an agreed report and reports on matters not so agreed, no later than 28 April 1994.

4. Proofs of evidence of witnesses of fact shall be exchanged no later than 2 weeks before the date of commencement of the Hearing.

5. The Claimant's opening address shall be submitted in writing no

later than one week before commencement of the Hearing. The Respondent's opening address, if needed, shall be submitted in writing on a date to be determined.

6. Arrangements for the Hearing are confirmed as follows:
   6.1 Commencement date: 1 June 1994.
   6.2 Period of time reserved: 20 working days.
   6.3 Venue: The Institution of Civil Engineers, Great George Street, London SW1P 3AA: the Claimant to reserve accommodation.
   6.4 Both parties have given notice of their intention to be represented at the Hearing by Junior Counsel. Any change of such intention must be notified to the other party and to me in sufficient time to allow for any resultant change that other party may wish to make in its representation.
   6.5 I shall make a tape recording of the Hearing: such recording being for my own use only as a supplement to my notes and not available to the parties unless I so decide; and to be erased after I have made my Final Award.

7. Should any further meetings be needed prior to the Hearing they will be held at the said Institution of Civil Engineers.

8. Both parties having so required I confirm that my Awards shall include reasons.

9. My Award on the substantive matters in dispute shall be in the form of an Interim Award. Thereafter an opportunity will be provided for the parties to address me, either orally or in writing as may be agreed, before I make my Final Award as to costs.

10. My Awards shall provide for taxation of the costs of the reference by me, pursuant to Section 18 of the Arbitration Act 1950, if such costs are not agreed.

11. Figures shall be agreed as figures where possible.

12. Plans, photographs and documents shall be agreed where possible.

13. The parties when writing to me shall send a copy of their letter and of any enclosures thereto to the other party, indicating to me that they have done so.

14. There shall be no communication by telephone between myself and either party. Urgent matters shall be communicated by fax.

15. I shall expect to receive copies of Pleadings, Lists of Documents, and of other communication relevant to applications or to matters referred to in these Directions; but not of routine communications between the parties where no action is required by me.

16. There shall be liberty to apply.

17. The costs of this Order shall be costs in the reference.

Dated this 22nd day of August 1993.

Arbitrator

To the Claimant's Representatives
To the Respondent's Representatives

SD/17 Points of Claim

IN THE MATTER OF THE ARBITRATION ACTS 1950–1979

AND

IN THE MATTER OF AN ARBITRATION BETWEEN

Majorspan Bridgebuilders PLC

CLAIMANT

AND

Universal Finance Corporation PLC

RESPONDENT

POINTS OF CLAIM

1. The Claimant is a civil engineering contractor based in Wessex. The Respondent is a banker providing finance for engineering projects.
2. By a contract under seal dated the 10th day of February 1991 ("the Contract") the Claimant undertook to construct and complete all permanent and temporary works in connection with a bridge over the River Avon, in consideration for certain payments to be made by the Respondent in accordance with the terms of the Contract.
3. The Contract incorporates a document commonly known as the ICE Conditions of Contract, 6th Edition, dated January 1991 ("the ICE Conditions").
4. Under Clause 12 of the ICE Conditions it is provided that the Contractor shall, subject to certain conditions, be paid the amount of any costs reasonably incurred by the Contractor by reason of physical conditions or artificial obstructions, together with a reasonable percentage addition thereto in respect of profit, where such conditions or obstructions could not reasonably have been foreseen by an experienced contractor.

5. Under Clause 60 (7) of the ICE Conditions it is provided that in the event of failure by the Engineer to certify or the Employer to make payment in accordance with the Contract the Employer shall pay to the Contractor interest compounded monthly at 2 per cent above base rate.
6. During the progress of the works the contractor encountered physical conditions and/or artificial obstructions; namely the presence of hard rock and concrete in ground required by the contract to be excavated; which conditions and/or obstructions could not reasonably have been foreseen by an experienced contractor.
7. As a consequence of having encountered the said physical conditions and/or artificial obstructions the Claimant incurred the costs of additional works and of delay to progress.
8. Contrary to the provisions of the said Clause 12 of the contract the Engineer failed to certify payment, and the Employer failed to pay, the additional costs and the profit addition thereto to which the Claimant is entitled in the sum of £3,824,522 plus interest thereon.

PARTICULARS

Additional Works and Delays

Plant costs	1,500,000
Labour costs	1,200,000
Material costs	600,000
Site overheads: 17 weeks at £18,120/week	308,040
Off-site overheads + profit at 6% of total	216,482
	£3,824,522

8. And accordingly the Claimant claims the sum of £3,824,522 plus interest thereon pursuant to Clause 60 (7) of the ICE Conditions at 2 per cent above base rate compounded monthly, plus costs.

Perry Mason

Served this Twentieth day of October 1993 by Newpastures & Co of 99 New Inn, Temple, Strand, London WC2; Solicitors for the Claimant.

SD/18 Request for Further and Better Particulars

IN THE MATTER OF THE ARBITRATION ACTS 1950-1979

AND

IN THE MATTER OF AN ARBITRATION BETWEEN

Majorspan Bridgebuilders PLC

CLAIMANT

AND

Universal Finance Corporation PLC

RESPONDENT

REQUEST FOR FURTHER AND BETTER PARTICULARS OF THE POINTS OF CLAIM

Under Paragraph 6

1. Please state the precise dates on which the Claimants allegedly encountered the physical conditions and/or artificial obstructions referred to.

Under Paragraph 8

1. Please provide full details of all costs claimed in respect of plant stating the type of each and every item, the period(s) during which it was in operation on the site, and the hire rate claimed.
2. Please provide full details of the items included under the heading "site overheads" including weekly or monthly rates claimed.

Barry Ster

Served this Second day of November 1993 by Callender & Co of Plastic Buildings, Gray's Inn, London; Solicitors for the Respondent.

SD/19 Further and Better Particulars

IN THE MATTER OF THE ARBITRATION ACTS 1950–1979

AND

IN THE MATTER OF AN ARBITRATION BETWEEN

Majorspan Bridgebuilders PLC — *CLAIMANT*

AND

Universal Finance Corporation PLC — *RESPONDENT*

FURTHER AND BETTER PARTICULARS OF THE POINTS OF CLAIM

Under Paragraph 6

1. Hard rock encountered on 02.04.91 and continued until 10.09.91. Concrete foundations encountered on 20.09.91 and continued until 10.10.91.

Under Paragraph 8

1. Details of plant costs are contained in computer print-outs which will be in evidence and will be made available on discovery and for agreement between the parties' experts.

2. Details of site overheads are given in the schedule appended hereto.

Perry Mason

Served this Sixteenth day of November 1993 by Newpastures & Co of 99 New Inn, Temple, Strand, London WC2; Solicitors for the Claimant.

SCHEDULE

SITE OVERHEADS	*PER WEEK*
Agent (including car and allowances)	650
Chief site engineer (ditto)	520
Junior site engineers: 5 no	2,000
Chief quantity surveyor	500
Junior QSs: 4 no	1,400
Office manager	450
Junior office staff: 4 no	1,400
Secretaries: 3 no	900
General foreman	500
Walking gangers: 6 no	2,400
Storekeeper	400
Timekeepers: 2 no	700
Fitters: 4 no	1,600
Hire of site offices	800
Cleaning services	400
Hire of workshop/stores	500
Contract insurances	1,800
Performance bond	1,200
TOTAL WEEKLY COST	**£18,120**

SD/20 Points of Defence and Counterclaim

IN THE MATTER OF THE ARBITRATION ACTS 1950-1979

AND

IN THE MATTER OF AN ARBITRATION BETWEEN

Majorspan Bridgebuilders PLC

CLAIMANT

AND

Universal Finance Corporation PLC

RESPONDENT

POINTS OF DEFENCE AND COUNTERCLAIM

POINTS OF DEFENCE

1. Paragraphs 1 and 2 of the Points of Claim are admitted.
2. Save that the Contract also incorporates certain additional and special clauses, together with a Specification, a Bill of Quantities and certain drawings, upon which the Respondent will rely for their full content and meaning, Paragraph 3 of the Points of Claim is admitted.
3. Save that the Respondent will rely upon the full content and meaning of the said Clauses 12 and 60 of the ICE Conditions, Paragraphs 4 and 5 of the Points of Claim are admitted.
4. It is admitted in respect of Paragraph 6 of the Points of Claim that the Claimant encountered certain physical conditions and/or artificial obstructions, namely the presence of hard rock and/or concrete during the progress of the works, and that the works were delayed thereby: but it is denied that the said conditions and/or obstructions could not reasonably have been foreseen by an experienced contractor.
5. It is averred that the said conditions and/or obstructions could and

should have been foreseen: and that by failing to do so the Claimant failed to exercise the skill reasonably to be expected of an experienced contractor.

6. The allegation in Paragraph 7 of the Points of Claim, that the Claimant incurred the costs of additional works and of delays, is not admitted. To the extent that such works were performed they ought to have been allowed for by the Claimant.

7. In respect of Paragraph 8 of the Points of Claim it is denied that the Engineer failed to certify, or that the Employer failed to pay, any sums to which the Claimant is entitled: and it is denied that the Claimant is entitled to be paid the sum of £3,824,522 or any other sum whatsoever.

8. Save as is herein expressly admitted the Respondent denies each and every allegation contained in the points of Claim as if the same were set out and specifically denied seriatim.

COUNTERCLAIM

9. The Claimant failed to complete the whole of the works within the Time for Completion referred to in the Contract: namely by 10 December 1992.

10. The whole of the works was not completed until 15 April 1993: being 18 weeks after the Date for Completion. The Employer is, under the terms of the Contract, entitled to liquidated damages at the rate of £80,000 for every week of delay.

11. The Respondent accordingly counterclaims the sum of £1,440,000 being its entitlement in respect of 18 weeks' delay.

Barry Ster

Served this Thirtieth day of December 1993 by Callender & Co of Plastic Buildings, Gray's Inn, London; Solicitors for the Respondent.

SD/21 Points of Reply and Defence to Counterclaim

IN THE MATTER OF THE ARBITRATION ACTS 1950–1979

AND

IN THE MATTER OF AN ARBITRATION BETWEEN

Majorspan Bridgebuilders PLC — *CLAIMANT*

AND

Universal Finance Corporation PLC — *RESPONDENT*

POINTS OF REPLY AND DEFENCE TO COUNTERCLAIM

POINTS OF REPLY

1. Save insofar as the Points of Defence consists of admissions the Claimant joins issue with the Respondent in each and every allegation contained therein.

2. It is denied that the physical conditions and artificial obstructions encountered by the Claimant could and should have been foreseen, as alleged in Paragraph 5 of the Points of Defence.

3. It is denied that the Claimant failed to exercise the skill reasonably to be expected of an experienced contractor, as alleged in Paragraph 5 of the Points of Claim.

4. It is denied that the additional works performed by the Claimant ought to have been allowed for.

DEFENCE TO COUNTERCLAIM

5. It is denied that the Claimant failed to complete the whole of the works within the Time for Completion referred to in the Contract, as alleged in Paragraph 9 of the Counterclaim.

6. In respect of Paragraph 10 of the Counterclaim, while it is admitted that the whole of the works was not completed until 15 April 1993: the cause of the delay, namely physical conditions and/or artificial obstructions, is such as fairly to entitle the Claimant to an extension of time under Clause 44 of the ICE Conditions. The Claimant applied for an extension of time of 18 weeks, in accordance with the said Clause 44, but the Engineer wrongly failed to grant such extension.

7. It is denied that the Respondent is entitled to £1,440,000 or to any sum whatsoever in respect of liquidated damages.

Perry Mason

Served this Fourteenth day of February 1994 by Newpastures & Co of 99 New Inn, Temple, Strand, London WC2; Solicitors for the Claimant.

SD/22 Scott Schedule

ITEM No	CLAIM	AMOUNT	DEFENCE	OFFER	ARBITRATOR'S FINDING	AWARD
1	Extension of time: inclement weather	6 weeks	Part of lost time foreseeable	2 weeks		
2	Extension of time: drwgs issued late	8 weeks	No delay caused	Nil		
3	Payment for ditto	£4,000	No liability	Nil		
4	Variations: founds: additional time	8 weeks	Admitted	8 weeks		
5	Ditto: additional excavation	£6,000	Quantity agreed. Rate reduced	£4500		
6	Ditto: alterations to sheet piling	£9,500	Piling to increased depth should have been allowed for	Nil		
7	Variations: general delay & disruption	20 weeks	Denied	Nil		
8	Ditto: additional labour costs	£8,700	Denied	Nil		
9	Ditto: additional site oncosts	£3,600	Denied	Nil		

SD/23 List of Documents

IN THE MATTER OF THE ARBITRATION ACTS 1950-1979

AND

IN THE MATTER OF AN ARBITRATION BETWEEN

Majorspan Bridgebuilders PLC
CLAIMANT

- AND -

Universal Finance Corporation PLC
RESPONDENT

CLAIMANT'S LIST OF DOCUMENTS

---

1. The schedules hereto are lists of the documents relating to the matters in question in this arbitration which are or have been in the possession custody or power of the above-named Claimant.

2. The Claimant has in his possession custody or power the documents relating to the matters in question in this arbitration enumerated in Schedule 1 hereto.

3. The Claimant objects to the production of the documents enumerated in Part 2 of the said Schedule 1 on the ground that they are documents written or obtained for the purpose of the Claimant's receiving legal advice or for the purpose of obtaining or furnishing information for material to be used as evidence on behalf of the Claimant.

4. The Claimant has had but does not now have in his possession custody or power the documents relating to the matters in question in this arbitration enumerated in Schedule 2 hereto.

5. The documents in the said Schedule 2 were last in the Claimant's possession custody or power on or about the dates thereof when they were dispatched to the persons to whom they were respectively addressed.

6. Neither the Claimant nor the Claimant's Solicitors nor any other person on his behalf has now or has ever had in his possession custody or power any document of any description whatsoever relating to any matter in question in this arbitration other than the documents enumerated in Schedules 1 and 2 hereto.

SCHEDULE 1 : PART 1

No	Description of Document	Date
1	Letter R to C inviting tender	01.08.90
2.	Copy letter C to R: confirm will tender	08.08.90
3.	Copy letter C to R enclosing tender	10.11.90
4.	Letter R to C: tender being considered	25.11.90
	Etc etc	

SCHEDULE 1 : PART 2

Description of Document

Correspondence passing between the Claimant, the Claimant's Solicitors, the Claimant's Counsel and Experts and notes and memoranda prepared by the Claimant and by his representatives and reports and statements of witnesses.

Various

SCHEDULE 2

Description of Document

Original letters written by the Claimant copies of which are specified in the First Schedule hereto.

Various

NOTE

In the above schedules the symbols used have the following meanings:

C	Claimant
R	Respondent
E	Engineer
RE	Engineer's Representative

NOTICE TO INSPECT

Take notice that the documents in the above list, other than those listed in Part 2 of Schedule 1 and in Schedule 2 may be inspected at the offices of the Claimant's Solicitors on any date by appointment.

SERVED this Twentieth day of March 1994 by Newpastures & Co of 99 New Inn, Temple, Strand, London WC2: Solicitors for the Claimant.

To the Respondent

Copy to the Arbitrator

SD/24 Calderbank letter

From: Universal Finance Corporation PLC

To: Majorspan Bridgebuilders PLC

16 March 1994

WITHOUT PREJUDICE SAVE AS TO COSTS

Dear Sirs

Arbitration between Majorspan Bridgebuilders PLC - and - Universal Finance Corporation PLC

Further to our informal and *without prejudice* discussion yesterday, at which your company was represented by Mr Brassey, Managing Director, and Mr Newpasture, solicitor, and my company was represented by Mr Sterling and its solicitor Mr Callender, I write to confirm the basis of our offer to settle this matter.

While my company remains convinced that its defence to your claim is entirely valid, it recognises that an early settlement would be beneficial to both parties in reducing uncertainty and the amount of costs likely to be incurred if the matter proceeds to a full hearing.

I am therefore authorised by my board to make the following offer:

1. That this company will, within fourteen days of your acceptance of this offer, pay to your company the sum of £2,000,000 (Two million pounds) in full and final settlement of all matters arising from this contract, and taking account both of claims and of counterclaims.

2. Each party shall bear its own costs of the reference.

3. Universal Finance Corporation will accept responsibility for the Arbitrator's fees and expenses.

This offer is made without admission of liability. If the offer is not accepted this letter will be privileged from discovery during the hearing of substantive issues: but it will in that event be adduced in evidence as a matter to be taken into consideration by the Arbitrator when making his award of costs.

Yours faithfully

H Appleby
Company Secretary

SD/25 Proof of Evidence

JOSEPH BLOGGS of 23 Railway Cuttings, East Cheam, Surrey, will say:

I have been employed by Majorspan Bridgebuilders PLC since 1975. I first joined that company as a carpenter and was engaged on a number of large building projects in London. In 1983 I was promoted to foreman carpenter and in 1988 I was again promoted, to general foreman. I worked in that capacity during 1991 and 1992 at the Avon Bridge site in Wessex.

On 2nd April 1991 excavation was in progress for the foundations to the west abutment, using a Hymac hydraulic digger which was loading the excavated material into lorries. The subsoil was mainly a sandy clay, and was being excavated without difficulty and in accordance with the working programme. However at about 11.00 am on that day hard obstructions were encountered, which prevented excavation from continuing in a large part of the abutment foundation area.

When the digger driver reported the problem to me I went with the agent, Mr Brassey, to the site of the abutment and examined the obstruction, which was hard rock at a level about 4 metres above formation level. As there had been no indication on any of the drawings, or in the borehole logs, that rock would be found we had no compressors or breakers on the site. I immediately ordered this equipment from our plant yard, and it arrived on site at 4.00 pm on that day, and was immediately put into operation.

It was clear that the excavation would take much longer than had been planned. I told the digger driver to excavate as much of the soft ground as he could reach while breakers were in use, and then to move on to the next foundation, at pier number 1.

However although it was possible to excavate down to a depth of about 2 metres at pier no 1, hard ground was again encountered. That excavation also had to be abandoned until breakers were available.

By this time it was clear that subsoil obstructions would be a major problem on this contract: mainly because rock had not been foreseen or allowed for in planning the work. Additional breakers were brought to the site to try to minimise the delay, and an additional digger was brought so that it could remove broken rock from foundations on the north side of the river, while the original digger worked on the south side.

Despite all of these measures excavation for the west abutment was not completed until the end of March 1992; about 5 weeks later than had been programmed. Other foundations, for piers numbered 1 and 2, and for the east abutment, were also delayed. The overall delay to the excavation as a result of the ground conditions was four months.

(Signed) Joseph Bloggs

SD/26 Interim Award

IN THE MATTER OF THE ARBITRATION ACTS 1950-1979

AND

IN THE MATTER OF AN ARBITRATION BETWEEN

Majorspan Bridgebuilders PLC

CLAIMANT

AND

Universal Finance Corporation PLC

RESPONDENT

I N T E R I M   A W A R D

WHEREAS:

1.1 By a Contract under seal dated the 10th day of February 1991 ("the Contract") the Claimant undertook to construct and complete certain works of civil engineering construction, namely a 3-span bridge over the River Avon, in consideration for which the Respondent undertook to pay to the Claimant the Contract Price at the times and in the manner prescribed by the Contract:

1.2 The Contract provided that any dispute between the parties that might arise from it should be referred to the arbitration of a person to be agreed upon between the parties; or, failing agreement, to be appointed upon the application of either party by the President or a Vice-President of the Institution of Civil Engineers:

1.3 A dispute having arisen and following upon an application by the Claimant on the 1st June 1993 the President of the said Institution did on the 20th July 1993 appoint me, Isambard Kingdom Brunel, Chartered Civil Engineer, of The Manse, Clifton, Bristol, to be Arbitrator in the reference: which appointment I accepted by notice in writing to both parties on the 25th July 1993:

1.4 The proposed terms of my appointment were set out in the Appendix to my letter dated 20th July 1993 and agreement thereto was confirmed by the Claimant's letter of 26th July 1993 and the Respondent's letter of 30th July 1993:

1.5 An Order for Directions was issued by me on the 22nd August 1993:

1.6 The said Order for Directions provided for an exchange of Pleadings and Discovery of Documents, followed by a Hearing in London commencing on 1st June 1994:

1.7 Following upon an exchange of Pleadings and Discovery of Documents the said Hearing at which the Claimant was represented by Perry Mason of Counsel and the Respondent by Barry Ster of Counsel commenced as arranged on 1st June 1994 and continued on working days thereafter until 15th June 1994:

1.8 The claim is for £3,824,522 plus interest thereon in respect of the costs of additional works and of delays resulting from the presence of certain physical conditions and/or artificial obstructions which the Claimants say could not reasonably have been foreseen by an experienced contractor:

1.9 The Respondent denies liability for the said costs on the ground that the said physical conditions and/or artificial obstructions were such as could reasonably have been foreseen by an experienced contractor: and the Respondent counterclaims £1,440,000 in respect of liquidated damages for delay in completion of the works:

1.10 The Claimant denies liability for the counterclaim on the ground that the period of admitted delay ought to have been the subject of an extension of time: which the Engineer refused to grant.

NOW I THE SAID ISAMBARD KINGDOM BRUNEL having heard and considered the evidence both oral and written adduced by both parties and the addresses to me by Counsel on their behalf, and having inspected real evidence both on the site and in the Respondent's testing laboratory DO HEREBY MAKE AND PUBLISH THIS MY INTERIM AWARD.

I FIND THAT:

2.1 The contract is for the construction of a 3-span road bridge over the River Avon. The substructure comprises two abutments, at the west and east ends of the bridge, and piers numbered 1 and 2.

2.2 Excavation for the foundations was delayed by the presence of hard solid rock which could not be removed with the Hymac excavator on site. The Claimant brought special equipment, namely compressors and breakers, to the site in order to deal with the obstructions.

2.3 Although the said special equipment was ultimately able to break out the several types of obstruction, progress of the works was delayed thereby, for a total of 17 weeks.

2.4 The costs of additional plant brought to the site in order to remove the obstructions, together with additional costs of plant, labour and site overheads resulting from the prolongation of the time for completion of the works, have been agreed between the parties, as figures, in the sum of £3,500,000.

2.5 In addition the parties have agreed upon an addition of 6 per cent of the above, namely £210,000, as being a reasonable evaluation of off-site overheads and profit.

2.6 Mr Ster, Counsel for the Respondent, while accepting that the figures referred to in paragraphs 2.4 and 2.5 above represent a fair evaluation of the costs referred to and of a reasonable addition for profit, submits that the claim is fundamentally flawed in that the obstructions encountered could and should have been foreseen.

2.7 Subject to that proviso, Mr Ster accepts on behalf of the Respondent that the total figure of £3,710,000 represents a reasonable evaluation of the costs and profit referred to in the claim.

2.8 The parties have further agreed that, if liability is established, the mean date on which payment ought to have been made may be taken, for purposes of interest calculations, as being 1st August 1991: and that, in that event, the counterclaim must automatically fail.

2.9 Mr Ster submits on behalf of the Respondent that the Site Investigation Report, the geological maps of the area of the site, and the presence of certain outcrops of rock near the site provide clear evidence of the probability that rock would be encountered during the excavations for the abutments and piers.

2.10 Mr Mason, Counsel for the Claimant, submits that no such inference may reasonably be drawn: and he urges me to accept both the validity in principle of the claim, and the Claimant's entitlement to such extension of time as will cover the delay that occurred.

2.11 Having carefully considered both learned Counsels' submissions, and after a detailed examination of the real and documentary evidence, I find that the obstructions encountered were not such as could reasonably have been foreseen by an experienced contractor.

2.12 I find that interest on the agreed evaluation of the claim, for the period from 1st August 1991 until 1st July 1994, calculated at 2 per cent above base rate and compounded monthly in accordance with Clause 60 (7) of the Contract, amounts to £1,545,596.

2.13 In addition I find that the delays caused by the presence of the physical obstructions are such as fairly to entitle the Claimant to an extension of time covering the period by which the works were delayed; namely 17 weeks.

I HOLD THAT:

3.1 The Respondent is liable to the Claimant in the sum of £3,710,000 in respect of additional costs and profit thereon, incurred in dealing with physical obstructions encountered, plus £1,545,596 in respect of interest thereon, making a total of £5,255,596:

3.2 The counterclaim, having regard to my findings of fact, is invalid.

AND ACCORDINGLY I HEREBY AWARD AND DIRECT THAT:

4.1 The Respondent shall within fourteen days of the date upon which this Award is taken up by either party pay to the Claimant the sum of £5,255,596 (five million, two hundred and fiftyfive thousand, five hundred and ninetysix pounds) in full and final settlement of all claims and counterclaims referred to me herein:

4.2 The costs of the reference shall be borne by such party or parties as I may direct in my Final Award, which will be made after I have heard both parties' submissions as to costs:

4.3 The costs of this my Interim Award, which costs I hereby tax and settle in the sum of £X plus Value Added Tax thereon of £Y, shall be borne by such party or parties as I may direct in my Final Award.

FIT FOR COUNSEL

Given under my hand this First day of July 1994

Arbitrator

In the presence of
Witness:
Address:
Occupation: Secretary

SD/27 Final Award

IN THE MATTER OF THE ARBITRATION ACTS 1950–1979

AND

IN THE MATTER OF AN ARBITRATION BETWEEN

Majorspan Bridgebuilders PLC

CLAIMANT

AND

Universal Finance Corporation PLC

RESPONDENT

F I N A L A W A R D

WHEREAS:

1.1 By a Contract under seal dated the 10th day of February 1991 ("the Contract") the Claimant undertook to construct and complete certain works of civil engineering construction, namely a 3-span bridge over the River Avon, in consideration for which the Respondent undertook to pay to the Claimant the Contract Price at the times and in the manner prescribed by the Contract:

1.2 The Contract provided that any dispute between the parties that might arise from it should be referred to the arbitration of a person to be agreed upon between the parties; or, failing agreement, to be appointed upon the application of either party by the President or a Vice-President of the Institution of Civil Engineers:

1.3 A dispute having arisen and following upon an application by the Claimant on the 1st June 1993 the President of the said Institution did on the 20th July 1993 appoint me, Isambard Kingdom Brunel, Chartered Civil Engineer, of The Manse, Clifton, Bristol, to be Arbitrator in the reference: which appointment I accepted by notice in writing to both parties on the 25th July 1993:

1.4 Following upon an exchange of pleadings, discovery of documents and a hearing I did on the 1st July 1994 make and publish an Interim Award wherein I determined the substantive issues in the reference, leaving costs to be determined in this my Final Award:

1.5 By letter dated 10th July 1994 I suggested to both parties that they might by agreement adopt a procedure whereby I would make my Final Award as to costs after receipt of written submissions thereon:

1.6 By letters dated the 15th and the 18th July 1994 the Claimant and the Respondent, respectively, agreed to my suggestion in paragraph 1.5 above:

1.7 By letter dated 20th July 1994 the Claimant submitted that, having succeeded in its claim, and the Respondent having failed in its counterclaim, the Claimant is entitled to be awarded its costs under the basic rule that *costs follow the event.*

1.8 By letter dated 27th July 1994 the Respondent submitted that its offer by way of its Calderbank letter dated 16 March 1994 was an invitation to negotiate a compromise: and that by failing to respond thereto the Claimant rejected that invitation. Accordingly the Respondent submits that responsibility for costs incurred subsequently to the date of the said Calderbank letter ought to be shared equally between the parties.

1.9 By letter dated 30th July 1994 the Claimant submitted that the Respondent's offer in settlement was inadequate and was not such as to warrant its acceptance or a departure from the rule that *costs follow the event.*

NOW I THE SAID ISAMBARD KINGDOM BRUNEL having considered the parties' written submissions DO HEREBY MAKE AND PUBLISH THIS MY FINAL AWARD.

I FIND AND HOLD THAT:

2.1 The Respondent's submission that its offer in settlement ought to affect my award of costs is without valid foundation:

2.2 There is no special reason why I ought to depart from the rule that *costs follow the event.*

AND ACCORDINGLY I HEREBY AWARD AND DIRECT THAT:

3.1 The Respondent shall pay and bear the Claimant's costs of the reference; such costs if not agreed to be taxed by me:

3.2 Within fourteen days of the date of this my Final Award the Respondent shall pay to the Claimant the sum of £X, being the amount of the costs of the award up to and including my Interim Award, already paid to me by the Claimant:

3.3 The costs of this my Final Award, which costs I hereby tax and settle in the sum of £Y plus Value Added Tax thereon of £Z, shall be borne by the Respondent: provided that, if such costs have already been paid to me by the Claimant, such costs shall within fourteen days of the date of this my Final Award be reimbursed to the Claimant by the Respondent.

FIT FOR COUNSEL

Given under my hand this Fifth day of August 1994

Arbitrator

In the presence of
Witness:
Address:
Occupation: Secretary

ppendix A

D/28 Notice of publication of Award

From: I K Brunel

1st July 1994

To: Majorspan Bridgebuilders PLC
Universal Finance Corporation PLC

Gentlemen

Arbitration between Majorspan Bridgebuilders PLC *(Claimant)* - and - Universal Finance Corporation PLC *(Respondent)*: Bridge over the River Avon

I have made and published my Interim Award in this reference and it is available for collection by or dispatch to the parties upon payment of my charges to date, which amount to £X including Value Added Tax.

The Award includes provision for reimbursement of those charges by the party responsible for them in the event that the other party pays the charges in taking up the Award.

Yours faithfully

Arbitrator

SD/29 Letter of Intent

1st February 1991

From: Universal Finance Corporation PLC

To: Majorspan Bridgebuilders PLC

Dear Sirs

<u>**Bridge over the River Avon**</u>

1. Subject to the granting of necessary planning consents it is the intention of Universal Finance Corporation ("the Corporation") to accept your tender dated 20th January 1991 for the above project.

2. It must be clearly understood that this letter does not constitute an acceptance of your tender. The Corporation does however hereby authorise you to proceed with the following preliminary works, subject to the right of the Corporation to rescind this authority at any time:
   2.1 The preparation of designs and drawings of temporary works
   2.2 The erection of site offices, workshops and stores
   2.3 The construction of temporary access roads
   2.4 The ordering of temporary and permanent works materials.

3. Payment for such work done and/or materials delivered to the site will be made on a *quantum meruit* basis subject to your compliance with all of the terms of this letter and of the tender documents, and subject to retention as provided for therein. Such payment will be credited against your entitlement to payment under any contract between the Corporation and yourselves that may come into being in respect of this project.

4. All goods and materials supplied and delivered to the site under the terms of this letter shall vest in the Corporation and shall be stored safely on the site until incorporated in the works or until the rescission of this authority. You shall provide insurances against all risks as specified in the tender documents while such goods and materials remain in your care and shall in the event of rescission of this authority forthwith deliver such goods and materials to the Corporation.

5. In the event that the Corporation exercises its power under paragraph 2 above to rescind this authority the Corporation shall within one month of such rescission pay to you all monies payable under the terms of this letter together with all retention monies held by the Corporation.

6. Any dispute or difference between the parties that may arise from this letter and your acceptance of its terms shall be referred to the arbitration of a person to be agreed upon or, failing agreement within one month of either party serving on the other a Notice to Concur, to be appointed by the President of the Institution of Civil Engineers.

7. Please confirm your acceptance of these terms by signing and returning the enclosed copy of this letter.

Yours faithfully

Appendix B: The Arbitration Act 1950 14 GEO. 6. CH. 27

(Showing amendments and deletions resulting from the Arbitration Act 1975, the Arbitration Act 1979, the Supreme Court Act 1981 and the Administration of Justice Act 1982.)

ARRANGEMENT OF SECTIONS

PART I
GENERAL PROVISIONS AS TO ARBITRATION

Effect of arbitration agreements, etc.

Section

Section *continued*

Provisions as to awards

13. Time for making award.
14. Interim awards.
15. Specific performance.
16. Awards to be final.
17. Power to correct slips.

Costs, fees and interests

18. Costs.
19. Taxation of arbitrator's or umpire's fees.
20. Interest on awards.

~~Special Cases,~~ Remission and setting aside of awards, etc. Repealed

21. ~~Statement of case.~~
22. Power to remit award.
23. Removal of arbitrator and setting aside of award.
24. Power of court to give relief where arbitrator is not impartial or the dispute involves question of fraud.
25. Power of court where arbitrator is removed or authority of arbitrator is revoked.

Enforcement of award

26. Enforcement of award.

Miscellaneous

27. Power of court to extend time for commencing arbitration proceedings.
28. Terms as to costs, etc.
29. Extension of s. 496 of the Merchant Shipping Act, 1894.
30. Crown to be bound.
31. Application of Part I to statutory arbitrations.
32. Meaning of 'arbitration agreement'.
33. Operation of Part I.
34. Extent of Part I.

PART II

ENFORCEMENT OF CERTAIN FOREIGN AWARDS

35. Awards to which Part II applies.
36. Effect of foreign awards.
37. Conditions for enforcement of foreign awards.
38. Evidence.
39. Meaning of 'final award'.
40. Saving for other rights, etc.

Section *continued*

CHAPTER 27

An Act to consolidate the Arbitration Act 1889 to 1934.
[28 July 1950]

Be it enacted by the King's most Excellent Majesty, by and with the advice and consent of the Lords Spiritual and Temporal, and Commons, in this present Parliament assembled, and by the authority of the same, as follows:

PART I
GENERAL PROVISIONS AS TO ARBITRATION

Effect of arbitration agreements, etc.

Authority of arbitrators and umpires to be irrevocable.

1. The authority of an arbitrator or umpire appointed by or by virtue of an arbitration agreement shall, unless a contrary intention is expressed in the agreement, be irrevocable except by leave of the High Court or a judge thereof.

Death of party.

2. (1) An arbitration agreement shall not be discharged by the death of any party thereto, either as respects the deceased or any other party, but shall in such an event be enforceable by or against the personal representative of the deceased.

(2) The authority of an arbitrator shall not be revoked by the death of any party by whom he was appointed.

Part I *continued*

(3) Nothing in this section shall be taken to affect the operation of any enactment or rule of law by virtue of which any right of action is extinguished by the death of a person.

Bankruptcy.

3. (1) Where it is provided by a term in a contract to which a bankrupt is a party that any differences arising thereout or in connection therewith shall be referred to arbitration, the said term shall, if the trustee in bankruptcy adopts the contract, be enforceable by or against him so far as relates to any such differences.

(2) Where a person who has been adjudged bankrupt had, before the commencement of the bankruptcy, become a party to an arbitration agreement, and any matter to which the agreement applies requires to be determined in connection with or for the purposes of the bankruptcy proceedings, then, if the case is one to which subsection (1) of this section does not apply, any other party to the agreement or, with the consent of the committee of inspection, the trustee in bankruptcy, may apply to the court having jurisdiction in the bankruptcy proceedings for an order directing that the matter in question shall be referred to arbitration in accordance with the agreement, and that court may, if it is of opinion that, having regard to all the circumstances of the case, the matter ought to be determined by arbitration, make an order accordingly.

Staying court proceedings where there is submission to arbitration.

4. (1) If any party to an arbitration agreement, or any person claiming through or under him, commences any legal proceedings in any court against any other party to the agreement, or any person claiming through or under him, in respect of any matter agreed to be referred, any party to those legal proceedings may at any time after appearance, and before delivering any pleadings or taking any other steps in the proceedings, apply to that court to stay the proceedings, and that court or a judge thereof, if satisfied that there is no sufficient reason why the matter should not be referred in accordance with the agreement, and that the applicant was, at the time when the proceedings were commenced, and still remains, ready and willing to do all things necessary to the proper conduct of the arbitration, may make an order staying the proceedings.

Part I *continued*

Repealed [1975 Act: s. 8(2)(a)].

~~(2) Notwithstanding anything in this Part of this Act, if any party to a submission to arbitration made in pursuance of an agreement to which the protocol set out in the First Schedule to this Act applies, or any person claiming through or under him, commences any legal proceedings in any court against any other party to the submission, or any person claiming through or under him, in respect of any matter agreed to be referred, any party to those legal proceedings may at any time after appearance, and before delivering any pleadings or taking any other steps in the proceedings, apply to that court to stay the proceedings, and that court or a judge thereof, unless satisfied that the agreement or arbitration has become inoperative or cannot proceed or that there is not in fact any dispute between the parties with regard to the matter agreed to be referred, shall make an order staying the proceedings.~~

Reference of interpleader issues to arbitration.

5. Where relief by way of interpleader is granted and it appears to the High Court that the claims in question are matters to which an arbitration agreement, to which the claimants are parties, applies, the High Court may direct the issue between the claimants to be determined in accordance with the agreement.

Arbitrators and umpires

When reference is to a single arbitrator.

6. Unless a contrary intention is expressed therein, every arbitration agreement shall, if no other mode of reference is provided, be deemed to include a provision that the reference shall be to a single arbitrator.

Power of parties in certain cases to supply vacancy.

7. Where an arbitration agreement provides that the reference shall be to two arbitrators, one to be appointed by each party, then, unless a contrary intention is expressed therein:

(a) if either of the appointed arbitrators refuses to act, or is incapable of acting, or dies, the party who appointed him may appoint a new arbitrator in his place;

(b) if, on such a reference, one party fails to appoint an arbitrator, either originally, or by way of substitution as aforesaid, for seven clear days after the other party, having appointed his arbitrator, has served the party making default with notice to make the appointment, the party who has appointed an

Part I *continued*

arbitrator may appoint that arbitrator to act as sole arbitrator in the reference and his award shall be binding on both parties as if he had been appointed by consent:

Provided that the High Court or a judge thereof may set aside any appointment made in pursuance of this section.

Umpires. As amended [1979 Act: s. 6(1)].

8. (1) Unless a contrary intention is expressed therein, every arbitration agreement shall, where the reference is to two arbitrators, be deemed to include a provision that the two arbitrators may appoint an umpire at any time after they are themselves appointed, and shall do so forthwith if they cannot agree.

(2) Unless a contrary intention is expressed therein, every arbitration agreement shall, where such a provision is applicable to the reference, be deemed to include a provision that if the arbitrators have delivered to any party to the arbitration agreement, or to the umpire, a notice in writing stating that they cannot agree, the umpire may forthwith enter on the reference in lieu of the arbitrators.

(3) At any time after the appointment of an umpire, however appointed the High Court may, on the application of any party to the reference and notwithstanding anything to the contrary in the arbitration agreement, order that the umpire shall enter upon the reference in lieu of the arbitrators and as if he were a sole arbitrator.

Majority award of three arbitrators as amended [1979 Act: s. 6(2)].

9. Unless a contrary intention is expressed in the arbitration agreement, in any case where there is a reference to three arbitrators, the award of any two of the arbitrators shall be binding.

Power of court in certain cases to appoint an arbitrator or umpire.

10. (1) In any of the following cases:

(a) where an arbitration agreement provides that the reference shall be to a single arbitrator, and all the parties do not, after differences have arisen, concur in the appointment of an arbitrator;

(b) if an appointed arbitrator refuses to act, or is incapable of acting, or dies, and the arbitration agreement does not show that it was intended that the vacancy should not be supplied and the parties do not supply the vacancy;

As amended [1979 Act: s. 6(3)].

(c) where the parties or two arbitrators are required or are at liberty to appoint an umpire or third arbitrator and do not appoint him;

Part I *continued*

(d) where an appointed umpire or third arbitrator refuses to act, or is incapable of acting, or dies, and the arbitration agreement does not show that it was intended that the vacancy should not be supplied, and the parties or arbitrators do not supply the vacancy;

any party may serve the other parties or the arbitrators, as the case may be, with a written notice to appoint or, as the case may be, concur in appointing, an arbitrator, umpire or third arbitrator, and if the appointment is not made within seven clear days after the service of the notice, the High Court or a judge thereof may, on application by the party who gave the notice, appoint an arbitrator, umpire or third arbitrator who shall have the like powers to act in the reference and make an award as if he had been appointed by consent of all parties.

As amended [1979 Act: s. 6(4)].

(2) In any case where:

(a) an arbitration agreement provides for the appointment of an arbitrator or umpire by a person who is neither one of the parties nor an existing arbitrator (whether the provision applies directly or in default of agreement by the parties or otherwise), and

(b) that person refuses to make the appointment or does not make it within the time specified in the agreement or, if no time is specified, within a reasonable time, any party to the agreement may serve the person in question with a written notice to appoint an arbitrator or umpire and, if the appointment is not made within seven clear days after service of the notice, the High Court or a judge thereof may, on the application of the party who gave the notice, appoint an arbitrator or umpire who shall have the like powers to act in the reference and make an award as if he had been appointed in accordance with the terms of the agreement.

As amended [C&LS Act 1990: s. 101(1)]. Section 10 of the 1950 Act continues to apply to arbitration agreements entered into before 1.4.91. Section 10(2) does not apply if a

(3) In any case where:

(a) an arbitration agreement provides that the reference shall be to three arbitrators, one to be appointed by each party and the third to be appointed by the two appointed by the parties or in some other manner specified in the agreement; and

(b) one of the parties ('the party in default') refuses to

Part I *continued* contrary intention is expressed in the arbitration agreement, whether or not as the result of a variation made after 1.4.91.

appoint an arbitrator or does not do so within the time specified in the agreement or, if no time is specified, within a reasonable time,

the other party to the agreement, having appointed his arbitrator, may serve the party in default with a written notice to appoint an arbitrator.

(3A) A notice under subsection (3) must indicate whether it is served for the purposes of subsection (3B) or for the purposes of subsection (3C).

(3B) Where a notice is served for the purposes of this subsection, then unless a contrary intention is expressed in the agreement, if the required appointment is not made within seven clear days after the service of the notice:

(a) the party who gave the notice may appoint his arbitrator to act as sole arbitrator in the reference; and

(b) his award shall be binding on both parties as if he had been appointed by consent.

(3C) Where a notice is served for the purposes of this subsection, then, if the required appointment is not made within seven clear days after the service of the notice, the High Court or a judge thereof may, on the application of the party who gave the notice, appoint an arbitrator on behalf of the party in default who shall have the like powers to act in the reference and make an award (and, if the case so requires, the like duty in relation to the appointment of a third arbitrator) as if he had been appointed in accordance with the terms of the agreement.

(3D) The High Court or a judge thereof may set aside any appointment made by virtue of subsection (3B).

Power of official referee to take arbitrations. As amended [C&LS Act 1990: s. 99].

11. (1) An official referee may, if in all the circumstances he thinks fit, accept appointment as sole arbitrator, or as umpire, by or by virtue of an arbitration agreement.

(2) An official referee shall not accept appointment as arbitrator or umpire unless the Lord Chief Justice has informed him that, having regard to the state of official referees' business, he can be made available to do so.

(3) The fees payable for the services of an official referee as arbitrator or umpire shall be taken in the High Court.

(4) Schedule 3 to the Administration of Justice Act 1970 1970 c. 31.

Part I *continued*

(which modifies this Act in relation to arbitration by judges, in particular by substituting the Court of Appeal for the High Court in provisions whereby arbitrators and umpires, their proceedings and awards are subject to control and review by the court) shall have effect in relation to official referees appointed as arbitrators or umpires as it has effect in relation to judge-arbitrators and judge-umpires (within the meaning of that Schedule).

(5) Any jurisdiction which is exercisable by the High Court in relation to arbitrators and umpires otherwise than under this Act shall, in relation to an official referee appointed as arbitrator or umpire, be exercisable instead by the Court of Appeal.

(6) In this section 'official referee' means any person nominated under section 68(1)(a) of the Supreme Court Act 1981 to deal with official referees' business.

1981 c. 54.

(7) Rules of the Supreme Court may make provision for:

(a) cases in which it is necessary to allocate references made under or by virtue of arbitration agreements to official referees;

(b) the transfer of references from one official referee to another.

Conduct of proceedings, witnesses, etc.

Conduct of proceedings, witnesses, etc.

12. (1) Unless a contrary intention is expressed therein, every arbitration agreement shall, where such a provision is applicable to the reference, be deemed to contain a provision that the parties to the reference, and all persons claiming through them respectively, shall, subject to any legal objection, submit to be examined by the arbitrator or umpire, an oath or affirmation, in relation to the matters in dispute, and shall, subject as aforesaid, produce before the arbitrator or umpire all documents within their possession or power respectively which may be required or called for, and do all other things which during the proceedings on the reference the arbitrator or umpire may require.

(2) Unless a contrary intention is expressed therein, every arbitration agreement shall, where such a provision is applicable to the reference, be deemed to contain a provision that the witnesses on the reference shall, if the arbitrator or umpire thinks fit, be examined on oath or affirmation.

(3) An arbitrator or umpire shall, unless a contrary intention is expressed in the arbitration agreement, have power to administer oaths to, or take the affirmation of, the parties to and witnesses on a reference under the agreement.

Part I *continued*

(4) Any party to a reference under an arbitration agreement may sue out a writ of subpoena *ad testificandum* or a writ of subpoena *duces tecum*, but no person shall be compelled under any such writ to produce any document which he could not be compelled to produce on the trial of an action, and the High Court or a judge thereof may order that a writ of subpoena *ad testificandum* or of subpoena *duces tecum* shall issue to compel the attendance before an arbitrator or umpire of a witness wherever he may be within the United Kingdom.

(5) The High Court or a judge thereof may also order that a writ of habeas corpus *ad testificandum* shall issue to bring up a prisoner for examination before an arbitrator or umpire.

(6) The High Court shall have, for the purpose of and in relation to a reference, the same power of making orders in respect of:

(a) security for costs;

~~(b) discovery of documents and interrogatories;~~

Repealed [C&LS Act 1990: s. 103].

(c) the giving of evidence by affidavit;

(d) examination on oath of any witness before an officer of the High Court or any other person, and the issue of a commission or request for the examination of a witness out of the jurisdiction;

(e) the preservation, interim custody or sale of any goods which are the subject matter of the reference;

(f) securing the amount in dispute in the reference;

(g) the detention, preservation or inspection of any property or thing which is the subject of the reference or as to which any question may arise therein, and authorizing for any of the purposes aforesaid any persons to enter upon or into any land or building in the possession of any party to the reference, or authorizing any samples to be taken or any observation to be made or experiment to be tried which may be necessary or expedient for the purpose of obtaining full information or evidence; and

Part I *continued*

(h) interim injunctions or the appointment of a receiver;

as it has for the purpose of and in relation to an action or matter in the High Court:

Provided that nothing in this subsection shall be taken to prejudice any power which may be vested in an arbitrator or umpire of making orders with respect to any of the matters aforesaid.

Provisions as to awards

Time for making award.

13. (1) Subject to the provisions of subsection (2) of section 22 of this Act, and anything to the contrary in the arbitration agreement, an arbitrator or umpire shall have power to make an award at any time.

(2) The time, if any, limited for making an award, whether under this Act or otherwise, may from time to time be enlarged by order of the High Court or a judge thereof, whether that time has expired or not.

(3) The High Court may, on the application of any party to a reference, remove an arbitrator or umpire who fails to use all reasonable dispatch in entering on and proceeding with the reference and making an award, and an arbitrator or umpire who is removed by the High Court under this subsection shall not be entitled to receive any remuneration in respect of his services.

For the purposes of this subsection, the expression 'proceeding with a reference' includes, in a case where two arbitrators are unable to agree, giving notice of that fact to the parties and to the umpire.

Want of prosecution. As amended [C&LS Act 1990: s. 102].

13A. (1) Unless a contrary intention is expressed in the arbitration agreement, the arbitrator or umpire shall have power to make an award dismissing any claim in a dispute referred to him if it appears to him that the conditions mentioned in subsection (2) are satisfied.

(2) The conditions are:

(a) that there has been inordinate and inexcusable delay on the part of the claimant in pursuing the claim; and

(b) that the delay:

(i) will give rise to a substantial risk that it is not possible to have a fair resolution of the issues in that claim; or

(ii) has caused, or is likely to cause or to have caused, serious prejudice to the respondent.

Part I *continued*

(3) For the purpose of keeping the provision made by this section and the corresponding provision which applies in relation to proceedings in the High Court in step, the Secretary of State may by order made by statutory instrument amend subsection (2) above.

(4) Before making any such order the Secretary of State shall consult the Lord Chancellor and such other persons as he considers appropriate.

(5) No such order shall be made unless a draft of the order has been laid before, and approved by resolution of, each House of Parliament.

Interim awards.

14. Unless a contrary intention is expressed therein, every arbitration agreement shall, where such a provision is applicable to the reference, be deemed to contain a provision that the arbitrator or umpire may, if he thinks fit, make an interim award, and any reference in this Part of this Act to an award includes a reference to an interim award.

Specific performance.

15. Unless a contrary intention is expressed therein, every arbitration agreement shall, where such a provision is applicable to the reference, be deemed to contain a provision that the arbitrator or umpire shall have the same power as the High Court to order specific performance of any contract other than a contract relating to land or any interest in land.

Awards to be final.

16. Unless a contrary intention is expressed therein, every arbitration agreement shall, where such a provision is applicable to the reference, be deemed to contain a provision that the award to be made by the arbitrator or umpire shall be final and binding on the parties and the persons claiming under them respectively.

Power to correct slips.

17. Unless a contrary intention is expressed in the arbitration agreement, the arbitrator or umpire shall have power to correct in an award any clerical mistake or error arising from any accidental slip or omission.

Costs, fees and interest

Costs.

18. (1) Unless a contrary intention is expressed therein, every arbitration agreement shall be deemed to include a provision that the costs of the reference and award shall be

Part I *continued*

in the discretion of the arbitrator or umpire, who may direct to and by whom and in what manner those costs or any part thereof shall be paid, and may tax or settle the amount of costs to be so paid or any part thereof, and may award costs to be paid as between solicitor and client.

(2) Any costs directed by an award to the paid shall, unless the award otherwise directs, be taxable in the High Court.

(3) Any provision in an arbitration agreement to the effect that the parties or any party thereto shall in any event pay their or his own costs of the reference or award or any part thereof shall be void, and this Part of this Act shall, in the case of an arbitration agreement containing any such provision, have effect as if that provision were not contained therein:

Provided that nothing in this subsection shall invalidate such a provision when it is a part of an agreement to submit to arbitration a dispute which has arisen before the making of that agreement.

(4) If no provision is made by an award with respect to the costs of the reference, any party to the reference may, within 14 days of the publication of the award or such further time as the High Court or a judge thereof may direct, apply to the arbitrator for an order directoring by and to whom those costs shall be paid, and thereupon the arbitrator shall, after hearing any party who may desire to be heard, amend his award by adding thereto such directions as he may think proper with respect to the payment of the costs of the reference.

(5) Section 69 of the Solicitors Act 1932 (which empowers a court before which any proceeding is being heard or is pending to charge property recovered or preserved in the proceeding with the payment of solicitors' costs) shall apply as if an arbitration were a proceeding in the High Court, and the High Court may make declarations and orders accordingly.

Taxation of arbitrator's or umpire's fees.

19. (1) If in any case an arbitrator or umpire refuses to deliver his award except on payment of the fees demanded by him, the High Court may, on an application for the purpose, order that the arbitrator or umpire shall deliver the award to the applicant on payment into court by the applicant of the fees demanded, and further that the fees

Part I *continued*

demanded shall be taxed by the taxing officer and that out of the money paid into court there shall be paid out to the arbitrator or umpire by way of fees such sum as may be found reasonable on taxation and that the balance of the money, if any, shall be paid out to the applicant.

(2) An application for the purposes of this section may be made by any party to the reference unless the fees demanded have been fixed by a written agreement between him and the arbitrator or umpire.

(3) A taxation of fees under this section may be reviewed in the same manner as a taxation of costs.

(4) The arbitrator or umpire shall be entitled to appear and be heard on any taxation or review of taxation under this section.

Power of arbitrator to award interest. Inserted under the Administration of Justice Act 1982: s. 15(6).

19A. (1) Unless a contrary intention is expressed therein, every arbitration agreement shall, where such a provision is applicable to the reference, be deemed to contain a provision that the arbitrator or umpire may, if he thinks fit, award simple interest at such rate as he thinks fit:

(a) on any sum which is the subject of the reference but which is paid before the award, for such period ending not later than the date of the payment as he thinks fit; and

(b) on any sum which he awards, for such period ending not later than the date of the award as he thinks fit.

(2) The power to award interest conferred on an arbitrator or umpire by subsection (1) above is without prejudice to any other power of an arbitrator or umpire to award interest.

Interest on awards.

20. A sum directed to be paid by an award shall, unless the award otherwise directs, carry interest as from the date of the award and at the same rate as a judgment debt.

~~Special cases~~, remission and setting aside of awards, etc.

~~Statement of case.~~ Repealed.

~~**21.** (1) An arbitrator or umpire may, and shall if~~ so ~~directed by the High Court, state –~~

~~(a) any question of law arising in the course of the reference; or~~

~~(b) an award or any part of an award,~~

~~in the form of a special case for the decision of the High Court.~~

Part I *continued*

~~(2) A special case with respect to an interim award or with respect to a question of law arising in the course of a reference may be stated, or may be directed by the High Court to be stated, notwithstanding that proceedings under the reference are still pending.~~

~~(3) A decision of the High Court under this section shall be deemed to be a judgment of the Court within the meaning of section 27 of the Supreme Court of Judicature (Consolidation) Act, 1925 (which relates to the jurisdiction of the Court of Appeal to hear and determine appeals from any judgment of the High Court), but no appeal shall lie from the decision of the High Court on any case stated under paragraph (a) of subsection (1) of this section without the leave of the High Court or of the Court of Appeal.~~

Power to remit award.

22. (1) In all cases of reference to arbitration the High Court or a judge thereof may from time to time remit the matters referred, or any of them, to the reconsideration of the arbitrator or umpire.

(2) Where an award is remitted, the arbitrator or umpire shall, unless the order otherwise directs, make his award within three months after the date of the order.

Removal of arbitrator and setting aside of award.

23. (1) Where an arbitrator or umpire has misconducted himself of the proceedings, the High Court may remove him.

(2) Where an arbitrator or umpire has misconducted himself or the proceedings, or an arbitration or award has been improperly procured, the High Court may set the award aside.

(3) Where an application is made to set aside an award, the High Court may order that any money made payable by the award shall be brought into court or otherwise secured pending the determination of the application.

Power of court to give relief where arbitrator is not impartial or the dispute involves question of fraud.

24. (1) Where an agreement between any parties provides that disputes which may arise in the future between them shall be referred to an arbitrator named or designated in the agreement, and after a dispute has arisen any party applies, on the ground that the arbitrator so named or designated is not or may not be impartial, for leave to revoke the authority of the arbitrator or for an injunction to restrain any other party or the arbitrator from proceeding with the arbitration, it shall not be a ground for refusing the application that the said party at the time when he made the agreement knew, or ought to have known, that the arbitrator, by reason of his relation towards any other party to the agreement or of his connection with the subject referred, might not be capable of impartiality.

Part I *continued*

(2) Where an agreement between any parties provides that disputes which may arise in the future between them shall be referred to arbitration, and a dispute which so arises involves the question whether any such party has been guilty of fraud, the High Court shall, so far as may be necessary to enable that question to be determined by the High Court, have power to order that the agreement shall cease to have effect and power to give leave to revoke the authority of any arbitrator or umpire appointed by or by virtue of the agreement.

(3) In any case where by virtue of this section the High Court has power to order that an arbitration agreement shall cease to have effect or to give leave to revoke the authority of an arbitrator or umpire, the High Court may refuse to stay any action brought in breach of the agreement.

Power of court to give relief where arbitrator is not impartial or the dispute involved question of fraud.

25. (1) Where an arbitrator (not being a sole arbitrator), or two or more arbitrators (not being all the arbitrators) or an umpire who has not entered on the reference is or are removed by the High Court or the Court of Appeal, the High Court may, on the application of any party to the arbitration agreement, appoint a person or persons to act as arbitrator or arbitrators or umpire in place of the person or persons so removed.

As amended [Administration of Justice Act 1970: S. 3].

(2) Where the authority of an arbitrator or arbitrators or umpire is revoked by leave of the High Court or the Court of Appeal, or a sole arbitrator or all the arbitrators or an umpire who has entered on the reference is or are removed by the High Court or the Court of Appeal, the High Court may, on the application of any party to the arbitration agreement, either:

(a) appoint a person to act as sole arbitrator in place of the person or persons removed; or

(b) order that the arbitration agreement shall cease to have effect with respect to the dispute referred.

(3) A person appointed under this section by the High

Part I *continued*

Court or the Court of Appeal, as an arbitrator or umpire shall have the like power to act in the reference and to make an award as if he had been appointed in accordance with the terms of the arbitration agreement.

(4) Where it is provided (whether by means of a provision in the arbitration agreement or otherwise) that an award under an arbitration agreement shall be a condition precedent to the bringing of an action with respect to any matter to which the agreement applies, the High Court or the Court of Appeal, if it orders (whether under this section or under any other enactment) that the agreement shall cease to have effect as regards any particular dispute, may further order that the provision making an award a condition precedent to the bringing of an action shall also cease to have effect as regards that dispute.

Enforcement of award

Enforcement of award.

26. (1) An award on an arbitration agreement may, by leave of the High Court or a judge thereof, be enforced in the same manner as a judgment or order to the same effect, and where leave is so given, judgment may be entered in terms of the award.

As amended [Administration of Justice Act 1977: s. 17(2)].

(2) If:

(a) the amount sought to be recovered does not exceed the current limit on jurisdiction in section 40 of the County Courts Act 1959, and

(b) a county court so orders,

it shall be recoverable (by execution issued from the county court or otherwise) as if payable under an order of that court and shall not be enforceable under subsection (1) above.

(3) An application to the High Court under this section shall preclude an application to a county court and an application to a county court under this section shall preclude an application to the High Court.

Miscellaneous

Power of court to extend time for commencing arbitration proceedings.

27. Where the terms of an agreement to refer future disputes to arbitration provide that any claims to which the agreement applies shall be barred unless notice to appoint an arbitrator is given or an arbitrator is appointed or some other step to commence arbitration proceedings is taken within a time fixed by the agreement, and a dispute arises to which the agreement applies, the High Court, if it is of opinion that in the circumstances of the case undue hardship would otherwise be caused, and notwithstanding that the time so fixed has expired, may, on such terms, if any, as the justice of the case may require, but without prejudice to the provisions of any enactment limiting the time for the commencement of arbitration proceedings, extend the time for such period as it thinks proper.

Part I *continued*

Terms as to costs, etc.

28. Any order made under this Part of the Act may be made on such terms as to costs or otherwise as the authority making the order thinks that:

~~Provided that this section shall not apply to any order made under subsection (2) of section 4 of this Act.~~

[As amended [1975 Act: s. 8(2)(b)].

Extension of s. 496 of the Merchant Shipping Act 1894.

29. (1) In subsection (3) of section 496 of the Merchant Shipping Act 1894 (which requires a sum deposited with a wharfinger by an owner of goods to be repaid unless legal proceedings are instituted by the shipowner), the expression 'legal proceedings' shall be deemed to include arbitration.

(2) For the purposes of the said section 496, as amended by this section, an arbitration shall be deemed to be commenced when one party to the arbitration agreement serves on the other party or parties a notice requiring him or them to appoint or concur in appointing an arbitrator, or, where the arbitration agreement provides that the reference shall be to a person named or designated in the agreement, requiring him or them to submit the dispute to the person so named or designated.

(3) Any such notice as is mentioned in subsection (2) of this section may be served either:

(a) by delivering it to the person on whom it is to be served; or

(b) by leaving it at the usual or last known place of abode in England of that person; or

(c) by sending it by post in a registered letter addressed to that person at his usual or last known place of abode in England;

as well as in any other manner provided in the arbitration agreement; and where a notice is sent by post in manner prescribed by paragraph (c) of this subsection, service thereof shall, unless the contrary is proved, be deemed to

Part I *continued*

have been effected at the time at which the letter would have been delivered in the ordinary course of post.

Crown to be bound. As amended [1975 Act: s. 8(2)(c)].

30. This Part of this Act ~~(except the provisions of subsection (2) of section 4 thereof)~~ shall apply to any arbitration to which His Majesty, either in right of the Crown or of the Duchy of Lancaster or otherwise, or the Duke of Cornwall, is a party.

Application of Part I to statutory arbitrations.

31. (1) Subject to the provisions of section 33 of this Act, this Part of this Act, except the provisions thereof specified in subsection (2) of this section, shall apply to every arbitration under any other Act (whether passed before or after the commencement of this Act) as if the arbitration were pursuant to an arbitration agreement and as if that other Act were an arbitration agreement, except in so far as this Act is inconsistent with that other Act or with any rules or procedure authorized or recognized thereby.

As amended [1975 Act: s. 8(2)(d)].

(2) The provisions referred to in subsection (1) of this section are subsection (1) of section 2, section 3, ~~subsection (2) of section 4,~~ section 5, subsection (3) of section 18 and sections 24, 25, 27 and 29.

Meaning of 'arbitration agreement'.

32. In this Part of this Act, unless the context otherwise requires, the expression 'arbitration agreement' means a written agreement to submit present or future differences to arbitration, whether an arbitrator is named therein or not.

Operation of Part I.

33. This Part of this Act shall not affect any arbitration commenced (within the meaning of subsection (2) of section 29 of this Act) before the commencement of this Act, but shall apply to an arbitration so commenced after the commencement of this Act under an agreement made before the commencement of this Act.

Extent of Part I. As amended [1975 Act: s. 8(2)(e)].

~~**34.** Subsection (2) of section 4 of this Act shall:~~

~~(a) extend to Scotland, with the omission of the words 'Notwithstanding anything in this Part of this Act' and with the substitution, for references to staying proceedings, of references to sisting proceedings; and~~

~~(b) extend to Northern Ireland, with the omission of the words 'Notwithstanding anything in this Part of this Act';~~

~~but, save as aforesaid.~~ None of the provisions of this Part of this Act shall extend to Scotland or Northern Ireland.

Part I *continued*

PART II
ENFORCEMENT OF CERTAIN FOREIGN AWARDS

Awards to which Part II applies.

35. (1) This Part of this Act applies to any award made after the 28th day of July, 1924:

(a) in pursuance of an agreement for arbitration to which the protocol set out in the First Schedule to this Act applies; and

(b) between persons of whom one is subject to the jurisdiction of some one of such Powers as His Majesty, being satisfied that reciprocal provisions have been made, may by Order in Council declare to be parties to the convention set out in the Second Schedule to this Act, and of whom the other is subject to the jurisdiction of some other of the Powers aforesaid; and

(c) in one of such territories as His Majesty, being satisfied that reciprocal provisions have been made, may by Order in Council declare to be territories to which the said convention applies;

and an award to which this Part of this Act applies is in this Part of this Act referred to as 'a foreign award'.

(2) His Majesty may by a subsequent Order in Council vary or revoke any Order previously made under this section.

(3) Any Order in Council under section one of the Arbitration (Foreign Awards) Act 1930, which is in force at the commencement of this Act shall have effect as if it had been made under this section.

Effect of foreign awards.

36. (1) A foreign award shall, subject to the provisions of this Part of this Act, be enforceable in England either by action or in the same manner as the award of an arbitrator is enforceable by virtue of section 26 of this Act.

(2) Any foreign award which would be enforceable under this Part of this Act shall be treated as binding for all purposes on the persons as between whom it was made, and may accordingly be relied on by any of those persons by way of defence, set off or otherwise in any legal proceedings in England, and any references in this Part of

Part II *continued*

this Act to enforcing a foreign award shall be construed as including references to relying on an award.

Conditions for enforcement of foreign awards.

37. (1) In order that a foreign award may be enforceable under this Part of this Act it must have:

(a) been made in pursuance of an agreement for arbitration which was valid under the law by which it was governed;

(b) been made by the tribunal provided for in the agreement or constituted in manner agreed upon by the parties;

(c) been made in conformity with the law governing the arbitration procedure;

(d) become final in the country in which it was made;

(e) been in respect of a matter which may lawfully be referred to arbitration under the law of England;

and the enforcement thereof must not be contrary to the public policy or the law of England.

(2) Subject to the provisions of this subsection, a foreign award shall not be enforceable under this Part of this Act if the court dealing with the case is satisfied that:

(a) the award has been annulled in the country in which it was made; or

(b) the party against whom it is sought to enforce the award was not given notice of the arbitration proceedings in sufficient time to enable him to present his case, or was under some legal incapacity and was not properly represented; or

(c) the award does not deal with all the questions referred or contains decisions on matters beyond the scope of the agreement for arbitration:

Provided that, if the award does not deal with all the questions referred, the court may, if it thinks fit, either postpone the enforcement of the award or order its enforcement subject to the giving of such security by the person seeking to enforce it as the court may think fit.

(3) If a party seeking to resist the enforcement of a foreign award proves that there is any ground other than the non-existence of the conditions specified in paragraphs (a), (b) and (c) of subsection (1) of this section, or the existence of the conditions specified in paragraphs (b) and (c) of subsection (2) of this section, entitling him to contest

the validity of the award, the court may, if it thinks fit, either refuse to enforce the award or adjourn the hearing until after the expiration of such period as appears to the court to be reasonably sufficient to enable that party to take the necessary steps to have the award annulled by the competent tribunal.

Part II *continued*

Evidence.

38. (1) The party seeking to enforce a foreign award must produce:

(a) the original award or a copy thereof duly authenticated in manner required by the law of the country in which it was made; and

(b) evidence proving that the award has become final; and

(c) such evidence as may be necessary to prove that the award is a foreign award and that the conditions mentioned in paragraphs (a), (b) and (c) of subsection (1) of the last foregoing section are satisfied.

(2) In any case where any document required to be produced under subsection (1) of this section is in a foreign language, it shall be the duty of the party seeking to enforce the award to produce a translation certified as correct by a diplomatic or consular agent of the country to which that party belongs, or certified as correct in such other manner as may be sufficient according to the law of England.

(3) Subject to the provisions of this section, rules of court may be made under section 84 of the Supreme Court Act 1981, with respect to the evidence which must be furnished by a party seeking to enforce an award under this part of this Act.

Meaning of 'final award'.

39. For the purposes of this Part of this Act, an award shall not be deemed final if any proceedings for the purpose of contesting the validity of the award are pending in the country in which it was made.

Saving for other rights, etc.

40. Nothing in this Part of this Act shall:

(a) prejudice any rights which any person would have had of enforcing in England any award or of availing himself in England of any award if neither this Part of this Act nor Part I of the Arbitration (Foreign Awards) Act 1930, had been enacted; or

Part II *continued*

(b) apply to any award made on an arbitration agreement governed by the law of England.

Application of Part II to Scotland.

41. (1) The following provisions of this section shall have effect for the purpose of the application of this Part of this Act to Scotland.

(2) For the references to England there shall be substituted references to Scotland.

(3) For subsection (1) of section 36 there shall be substituted the following subsection:

'(1) A foreign award shall, subject to the provisions of this Part of this Act, be enforceable by action, or, if the agreement for arbitration contains consent to the registration of the award in the Books of Council and Session for execution and the award is so registered, it shall, subject as aforesaid, be enforceable by summary diligence.'

(4) For subsection (3) of section 38 there shall be substituted the following subsection:

'(3) The Court of Session shall, subject to the provisions of this section, have power, exercisable by statutory instrument, to make provision by Act of Sederunt with respect to the evidence which must be furnished by a party seeking to enforce in Scotland an award under this Part of this Act, and the Statutory Instruments Act 1946, shall apply to a statutory instrument containing an Act of Sederunt made under this subsection as if the Act of Sederunt had been made by a Minister of the Crown.'

Application of Part II to Northern Ireland.

42. (1) The following provisions of this section shall have effect for the purpose of the application of this Part of this Act to Northern Ireland.

(2) For the references to England there shall be substituted references to Northern Ireland.

(3) For subsection (1) of section 36 there shall be substituted the following subsection:

'(1) A foreign award shall, subject to the provisions of this Part of this Act, be enforceable either by action or in the same manner as the award of an arbitrator under the provisions of the Common Law Procedure Amendment Act (Ireland) 1856, was enforceable at the date of the passing of the Arbitration (Foreign Awards) Act 1930.'

Part II *continued*

(4) For the reference, in subsection (3) of section 38 to section 99 of the Supreme Court of Judicature (Consolidation) Act 1925, there shall be substituted a reference to section 61 of the Supreme Court of Judicature (Ireland) Act 1877, as amended by any subsequent enactment.

Saving for pending proceedings.

43. Any proceedings instituted under Part I of the Arbitration (Foreign Awards) Act 1930, which are uncompleted at the commencement of this Act may be carried on and completed under this Part of this Act as if they had been instituted thereunder.

PART III
GENERAL

Short title, commencement and repeal.

44. (1) This Act may be cited as the Arbitration Act 1950.

(2) This Act shall come into operation on the first day of September, nineteen hundred and fifty.

(3) The Arbitration Act 1889, the Arbitration Clauses (Protocol) Act 1924, and the Arbitration Act 1934, are hereby repealed except in relation to arbitrations commenced (within the meaning of subsection (2) of section 29 of this Act) before the commencement of this Act, and the Arbitration (Foreign Awards) Act 1930, is hereby repealed; and any reference in any Act or other document to any enactment hereby repealed shall be construed as including a reference to the corresponding provision of this Act.

SCHEDULES

FIRST SCHEDULE

Sections 4, 35.

PROTOCOL ON ARBITRATION CLAUSES SIGNED ON BEHALF OF HIS MAJESTY AT A MEETING OF THE ASSEMBLY OF THE LEAGUE OF NATIONS HELD ON THE TWENTY-FOURTH DAY OF SEPTEMBER, NINETEEN HUNDRED AND TWENTY-THREE

The undersigned, being duly authorized, declare that they accept, on behalf of the countries which they represent, the following provisions:

First Schedule *continued*

1. Each of the Contracting States recognizes the validity of an agreement whether relating to existing or future differences between parties, subject respectively to the jurisdiction of different Contracting States by which the parties to a contract agree to submit to arbitration all or any differences that may arise in connection with such contract relating to commercial matters or to any other matter capable of settlement by arbitration, whether or not the arbitration is to take place in a country to whose jurisdiction none of the parties is subject.

Each Contracting State reserves the right to limit the obligation mentioned above to contracts which are considered as commercial under its national law. Any Contracting State which avails itself of this right will notify the Secretary-General of the League of Nations, in order that the other Contracting States may be so informed.

2. The arbitral procedure, including the constitution of the arbitral tribunal, shall be governed by the will of the parties and by the law of the country in whose territory the arbitration takes place.

The Contracting States agree to facilitate all steps in the procedure which require to be taken in their own territories, in accordance with the provisions of their law governing arbitral procedure applicable to existing differences.

3. Each Contracting State undertakes to ensure the execution by its authorities and in accordance with the provisions of its national laws of arbitral awards made in its own territory under the preceding articles.

4. The tribunals of the Contracting Parties, on being seized of a dispute regarding a contract made between persons to whom Article 1 applies and including an arbitration agreement whether referring to present or future differences which is valid in virtue of the said article and capable of being carried into effect, shall refer the parties on the application of either of them to the decision of the arbitrators.

Such reference shall not prejudice the competence of the judicial tribunals in case the agreement or the arbitration cannot proceed or become inoperative.

5. The present Protocol, which shall remain open for signature by all States, shall be ratified. The ratifications

First Schedule *continued*

shall be deposited as soon as possible with the Secretary-General of the League of Nations, who shall notify such deposit to all the signatory States.

6. The present Protocol shall come into force as soon as two ratifications have been deposited. Thereafter it will take effect, in the case of each Contracting State, one month after the notification by the Secretary-General of the deposit of its ratification.

7. The present Protocol may be denounced by any Contracting State on giving one year's notice. Denunciation shall be effected by a notification addressed to the Secretary-General of the League, who will immediately transmit copies of such notification to all the other signatory States and inform them of the date of which it was received. The denunciation shall take effect one year after the date on which it was notified to the Secretary-General, and shall operate only in respect of the notifying State.

8. The Contracting States may declare that their acceptance of the present Protocol does not include any or all of the under-mentioned territories: that is to say, their colonies, overseas possessions or territories, protectorates or the territories over which they exercise a mandate.

The said States may subsequently adhere separately on behalf of any territory thus excluded. The Secretary-General of the League of Nations shall be informed as soon as possible of such adhesions. He shall notify such adhesions to all signatory States. They will take effect one month after the notification by the Secretary-General to all signatory States.

The Contracting States may also denounce the Protocol separately on behalf of any of the territories referred to above. Article 7 applies to such denunciation.

SECOND SCHEDULE

Section 35.

CONVENTION ON THE EXECUTION OF FOREIGN ARBITRAL AWARDS SIGNED AT GENEVA ON BEHALF OF HIS MAJESTY ON THE TWENTY-SIXTH DAY OF SEPTEMBER, NINETEEN HUNDRED AND TWENTY-SEVEN

Second Schedule *continued*

ARTICLE 1

In the territories of any High Contracting Party to which the present Convention applies, an arbitral award made in pursuance of an agreement, whether relating to existing or future differences (hereinafter called 'a submission to arbitration') covered by the Protocol on Arbitration Clauses, opened at Geneva on 24 September 1923, shall be recognized as binding and shall be enforced in accordance with the rules of the procedure of the territory where the award is relied upon, provided that the said award has been made in a territory of one of the High Contracting Parties to which the present Convention applies and between persons who are subject to the jurisdiction of one of the High Contracting Parties.

To obtain such recognition or enforcement, it shall, further, be necessary:

(a) that the award has been made in pursuance of a submission to arbitration which is valid under the law applicable thereto;

(b) that the subject-matter of the award is capable of settlement by arbitration under the law of the country in which the award is sought to be relied upon;

(c) that the award has been made by the Arbitral Tribunal provided for in the submission to arbitration or constituted in the manner agreed upon by the parties and in conformity with the law governing the arbitration procedure;

(d) that the award has become final in the country in which it has been made, in the sense that it will not be considered as such if it is open to *opposition*, *appel* or *pourvoi en cassation* (in the countries where such forms of procedure exist) or if it is proved that any proceedings for the purpose of contesting the validity of the award are pending;

(e) that the recognition or enforcement of the award is not contrary to the public policy or to the principles of the law of the country in which it is sought to be relied upon.

ARTICLE 2

Even if the conditions laid down in Article 1 hereof are

fulfilled, recognition and enforcement of the award shall be refused if the Court is satisfied:

Second Schedule *continued*

(a) that the award has been annulled in the country in which it was made;

(b) that the party against whom it is sought to use the award was not given notice of the arbitration proceedings in sufficient time to enable him to present his case; or that, being under a legal incapacity, he was not properly represented;

(c) that the award does not deal with the differences contemplated by or falling within the terms of the submission to arbitration or that it contains decisions on matters beyond the scope of the submission to arbitration.

If the award has not covered all the questions submitted to the arbitral tribunal, the competent authority of the country where recognition or enforcement of the award is sought can, if it think fit, postpone such recognition or enforcement or grant it subject to such guarantee as that authority may decide.

ARTICLE 3

If the party against whom the award has been made proves that, under the law governing the arbitration procedure, there is a ground, other than the grounds referred to in Article 1(a) and (c), and Article 2(b) and (c), entitling him to contest the validity of the award in a Court of Law, the Court may, if it thinks fit, either refuse recognition or enforcement of the award or adjourn the consideration thereof, giving such party a reasonable time within which to have the award annulled by the competent tribunal.

ARTICLE 4

The party relying upon an award or claiming its enforcement must supply, in particular:

(1) The original award or a copy thereof duly authenticated, according to the requirements of the law of the country in which it was made;

(2) Documentary or other evidence to prove that the award has become final in the sense defined in

Article 1(d), in the country in which it was made;

(3) When necessary, documentary or other evidence to prove that the conditions laid down in Article 1, paragraph 1 and paragraph 2(a) and (c), have been fulfilled.

A translation of the award and of the other documents mentioned in this Article into the official language of the country where the award is sought to be relied upon may be demanded. Such translation must be certified correct by a diplomatic or consular agent of the country to which the party who seeks to rely upon the award belongs or by a sworn translator of the country where the award is sought to be relied upon.

ARTICLE 5

The provisions of the above Articles shall not deprive any interested party of the right of availing himself of an arbitral award in the manner and to the extent allowed by the law or the treaties of the country where such award is sought to be relied upon.

ARTICLE 6

The present Convention applies only to arbitral awards made after the coming into force of the Protocol on Arbitration Clauses, opened at Geneva on 24 September 1923.

ARTICLE 7

The present Convention, which will remain open to the signature of all the signatories of the Protocol of 1923 on Arbitration Clauses, shall be ratified.

It may be ratified only on behalf of those Members of the League of Nations and non-Member States on whose behalf the Protocol of 1923 shall have been ratified.

Ratifications shall be deposited as soon as possible with the Secretary-General of the League of Nations, who will notify such deposit to all the signatories.

ARTICLE 8

The present Convention shall come into force three

months after it shall have been ratified on behalf of two High Contracting Parties. Thereafter, it shall take effect, in the case of each High Contracting Party, three months after the deposit of the ratification on its behalf with the Secretary-General of the League of Nations.

Article 9

The present Convention may be denounced on behalf of any Member of the League or non-Member State. Denunciation shall be notified in writing to the Secretary-General of the League of Nations, who will immediately send a copy thereof, certified to be in conformity with the notification, to all the other Contracting Parties, at the same time informing them of the date on which he received it.

The denunciation shall come into force only in respect of the High Contracting Party which shall have notified it and one year after such notification shall have reached the Secretary-General of the League of Nations.

The denunciation of the Protocol on Arbitration Clauses shall entail, ipso facto, the denunciation of the present Convention.

Article 10

The present Convention does not apply to the Colonies, Protectorates or territories under suzerainty or mandate of any High Contracting Party unless they are specially mentioned.

The application of this Convention to one or more of such Colonies, Protectorates or territories to which the Protocol on Arbitration Clauses, opened at Geneva on 24 September 1923, applies, can be effected at any time by means of a declaration addressed to the Secretary-General of the League of Nations by one of the High Contracting Parties.

Such declaration shall take effect three months after the deposit thereof.

The High Contracting Parties can at any time denounce the Convention for all or any of the Colonies, Protectorates or territories referred to above. Article 9 hereof applies to such denunciation.

Second Schedule *continued*

ARTICLE 11

A certified copy of the present Convention shall be transmitted by the Secretary-General of the League of Nations to every Member of the League of Nations and to every non-Member State which signs the same.

Table of Statutes referred to in this Act

Short title	Session and Chapter
Common Law Procedure Amendment Act (Ireland) 1856	19 & 20 Vict. c. 102.
Supreme Court of Judicature (Ireland) Act 1877	40 & 41 Vict. c. 57.
Arbitration Act 1889	52 & 53 Vict. c. 49.
Merchant Shipping Act 1894	57 & 58 Vict. c. 60.
Arbitration Clauses (Protocol) Act 1924 ..	14 & 15 Geo. 5. c. 39.
Supreme Court of Judicature (Consolidation) Act 1925	15 & 16 Geo. 5. c. 49.
Arbitration (Foreign Awards) Act 1930 ..	20 Geo. 5. c. 15.
Arbitration Act 1934	24 & 25 Geo. 5. c. 14.
Statutory Instruments Act 1946	9 & 10 Geo. 6. c. 36.

Appendix C: Arbitration Act 1975

1975 CHAPTER 3

An Act to give effect to the New York Convention on the Recognition and Enforcement of Foreign Arbitral Awards.
[25 February 1975]

Be it enacted by the Queen's most Excellent Majesty, by and with the advice and consent of the Lords Spiritual and Temporal, and Commons, in this present Parliament assembled, and by the authority of the same, as follows:

Effect of arbitration agreement on court proceedings

Staying court proceedings where party proves arbitration agreement.

1. (1) If any party to an arbitration agreement to which this section applies, or any person claiming through or under him, commences any legal proceedings in any court against any other party to the agreement, or any person claiming through or under him, in respect of any matter agreed to be referred, any party to the proceedings may at any time after appearance, and before delivering any pleadings or taking any other steps in the proceedings, apply to the court to stay the proceedings; and the court, unless satisfied that the arbitration agreement is null and void, inoperative or incapable of being performed or that there is not in fact any dispute between the parties with regard to the matter agreed to be referred, shall make an order staying the proceedings.

(2) This section applies to any arbitration agreement which is not a domestic arbitration agreement; and neither section 4(1) of the Arbitration Act 1950 nor section 4 of the Arbitration Act (Northern Ireland) 1937 shall apply to an arbitration agreement to which this section applies.

1950 c. 27.
1937 c. 8 (N.I.).

(3) In the application of this section to Scotland, for the references to staying proceedings there shall be substituted references to sisting proceedings.

(4) In this section 'domestic arbitration agreement' means an arbitration agreement which does not provide, expressly or by implication, for arbitration in a State other than the United Kingdom and to which neither:

(a) an individual who is a national of, or habitually resident in, any State other than the United Kingdom; nor

(b) a body corporate which is incorporated in, or whose central management and control is exercised in, any State other than the United Kingdom;

is a party at the time the proceedings are commenced.

Enforcement of Convention awards

Replacement of former provisions. 1950 c. 27.

2. Sections 3 to 6 of this Act shall have effect with respect to the enforcement of Convention awards; and where a Convention award would, but for this section, be also a foreign award within the meaning of Part II of the Arbitration Act 1950, that party shall not apply to it.

Effect of Convention awards.

3. (1) A Convention award shall, subject to the following provisions of this Act, be enforceable:

(a) in England and Wales, either by action or in the same manner as the award of an arbitrator is enforceable by virtue of section 26 of the Arbitration Act 1950;

(b) in Scotland, either by action or, in a case where the arbitration agreement contains consent to the registration of the award in the Books of Council and Session for execution and the award is so registered, by summary diligence;

(c) in Northern Ireland, either by action or in the same manner as the award of an arbitrator is enforceable by virtue of section 16 of the Arbitration Act (Northern Ireland) 1937.

1937 c. 8 (N.I.).

(2) Any Convention award which would be enforceable under this Act shall be treated as binding for all purposes on the persons as between whom it was made, and may accordingly be relied on by any of those persons by way of defence, set off or otherwise in any legal proceedings in the United Kingdom; and any reference in this Act to

enforcing a Convention award shall be construed as including references to relying on such an award.

Evidence.

4. The party seeking to enforce a Convention award must produce:

(a) the duly authenticated original award or a duly certified copy of it; and

(b) the original arbitration agreement or a duly certified copy of it; and

(c) where the award or agreement is in a foreign language, a translation of it certified by an official or sworn translator or by a diplomatic or consular agent.

Refusal of enforcement.

5. (1) Enforcement of a Convention award shall not be refused except in the cases mentioned in this section.

(2) Enforcement of a Convention award may be refused if the person against whom it is invoked proves:

(a) that a party to the arbitration agreement was (under the law applicable to him) under some incapacity; or

(b) that the arbitration agreement was not valid under the law to which the parties subjected it or, failing any indication thereon, under the law of the country where the award was made; or

(c) that he was not given proper notice of the appointment of the arbitrator or of the arbitration proceedings or was otherwise unable to present his case; or

(d) (subject to subsection (4) of this section) that the award deals with a difference not contemplated by or not falling within the terms of the submission to arbitration or contains decisions on matters beyond the scope of the submission to arbitration; or

(e) that the composition of the arbitral authority or the arbitral procedure was not in accordance with the agreement of the parties or, failing such agreement, with the law of the country where the arbitration took place; or

(f) that the award has not yet become binding on the parties, or has been set aside or suspended by a competent authority of the country in which, or under the law of which, it was made.

(3) Enforcement of a Convention award may also be

refused if the award is in respect of a matter which is not capable of settlement by arbitration, or if it would be contrary to public policy to enforce the award.

(4) A Convention award which contains decisions on matters not submitted to arbitration may be enforced to the extent that it contains decisions on matters submitted to arbitration which can be separated from those on matters not so submitted.

(5) Where an application for the setting aside or suspension of a Convention award has been made to such a competent authority as is mentioned in subsection (2)(f) of this section, the court before which enforcement of the award is sought may, if it thinks fit, adjourn the proceedings and may, on the application of the party seeking to enforce the award, order the other party to give security.

Saving.

1950 c. 27.

6. Nothing in this Act shall prejudice any right to enforce or rely on an award otherwise than under this Act or Part II of the Arbitration Act 1950.

General

Interpretation.

7. (1) In this Act:

'arbitration agreement' means an agreement in writing (including an agreement contained in an exchange of letters or telegrams) to submit to arbitration present or future differences capable of settlement by arbitration;

'Convention award' means an award made in pursuance of an arbitration agreement in the territory of a State, other than the United Kingdom, which is a party to the New York Convention; and

'the New York Convention' means the Convention on the Recognition and Enforcement of Foreign Arbitral Awards adopted by the United Nations Conference on International Commercial Arbitration on 10 June 1958.

(2) If Her Majesty by Order in Council declares that any State specified in the Order is a party to the New York Convention the Order shall, while in force, be conclusive evidence that that State is a party to that Convention.

(3) An Order in Council under this section may be varied or revoked by a subsequent Order in Council.

Short title, repeals, commencement and extent.

8. (1) This Act may be cited as the Arbitration Act 1975.

(2) The following provisions of the Arbitration Act 1950 are hereby repealed, that is to say:

(a) section 4(2);

(b) in section 28 the proviso;

(c) in section 30 the words '(except the provisions of subsection (2) of section 4 thereof)';

(d) in section 31(2) the words 'subsection (2) of section 4'; and

(e) in section 34 the words from the beginning to 'save as aforesaid'.

(3) This Act shall come into operation on such date as the Secretary of State may by order made by statutory instrument appoint.

(4) This Act extends to Northern Ireland.

Appendix D: Arbitration Act 1979

(Showing amendments resulting from the Supreme Court Act 1981: which amendments shall not have effect as regards decisions of the High Court pronounced before 1 January 1982.)

1979 CHAPTER 42

An Act to amend the law relating to arbitrations and for purposes connected therewith.

[4 April 1979]

Be it enacted by the Queen's most Excellent Majesty, by and with the advice and consent of the Lords Spiritual and Temporal, and Commons, in this present Parliament assembled, and by the authority of the same, as follows:

Judicial review of arbitration awards. 1950 c. 27.

1. (1) In the Arbitration Act 1950 (in this Act referred to as 'the principal Act') section 21 (statement of case for a decision of the High Court) shall cease to have effect and, without prejudice to the right of appeal conferred by subsection (2) below, the High Court shall not have jurisdiction to set aside or remit an award on an arbitration agreement on the ground of errors of fact or law on the face of the award.

(2) Subject to subsection (3) below, an appeal shall lie to the High Court on any question of law arising out of an award made on an arbitration agreement; and on the determination of such an appeal the High Court may by order

(a) confirm, vary or set aside the award; or

(b) remit the award to the reconsideration of the arbitrator or umpire together with the court's opinion on the question of law which was the subject of the appeal;

and where the award is remitted under paragraph (b) above the arbitrator or umpire shall, unless the order otherwise directs, make his award within three months after the date of the order.

(3) An appeal under this section may be brought by any of the parties to the reference

(a) with the consent of all the other parties to the reference; or

(b) subject to section 3 below, with the leave of the court.

(4) The High Court shall not grant leave under subsection (3)(b) above unless it considers that, having regard to all the circumstances, the determination of the question of law concerned could substantially affect the rights of one or more of the parties to the arbitration agreement; and the court may make any leave which it gives conditional upon the applicant complying with such conditions as it considers appropriate.

(5) Subject to subsection (6) below, if an award is made and, on an application made by any of the parties to the reference

(a) with the consent of all the other parties to the reference, or

(b) subject to section 3 below, with the leave of the court,

it appears to the High Court that the award does not or does not sufficiently set out the reasons for the award, the court may order the arbitrator or umpire concerned to state the reasons for his award in sufficient detail to enable the court, should an appeal be brought under this section, to consider any question of law arising out of the award.

(6) In any case where an award is made without any reason being given, the High Court shall not make an order under subsection (5) above unless it is satified

(a) that before the award was made one of the parties to the reference gave notice to the arbitrator or umpire concerned that a reasoned award would be required; or

(b) that there is some special reason why such a notice was not given.

Amended under the Supreme Court Act 1981 s. 148(2).

(6A) Unless the High Court gives leave, no appeal shall lie to the Court of Appeal from a decision of the High Court

(a) to grant or refuse leave under subsection (3)(b) or (5)(b) above; or

(b) to make or not to make an order under subsection (5) above.

(7) No appeal shall lie to the Court of Appeal from a decision of the High Court on an appeal under this section unless

(a) the High Court or the Court of Appeal gives leave; and

(b) it is certified by the High Court that the question of law to which its decision relates either is one of general public importance or is one which for some other special reason should be considered by the Court of Appeal.

(8) Where the award of an arbitrator or umpire is varied on appeal, the award as varied shall have effect (except for the purposes of this section) as if it were the award of the arbitrator or umpire.

Determination of preliminary point of law by court.

2. (1) Subject to subsection (2) and section 3 below, on an application to the High Court made by any of the parties to a reference:

(a) with the consent of an arbitrator who has entered on the reference or, if an umpire has entered on the reference, with his consent, or

(b) with the consent of all the other parties,

the High Court shall have jurisdiction to determine any question of law arising in the course of the reference.

(2) The High Court shall not entertain an application under subsection (1)(a) above with respect to any question of law unless it is satisfied that:

(a) the determination of the application might produce substantial savings in costs to the parties; and

(b) the question of law is one in respect of which leave to appeal would be likely to be given under section 1(3)(b) above.

'(2A) Unless the High Court gives leave, no appeal shall lie to the Court of Appeal from a decision of the High Court to entertain or not to entertain an application under subsection (1)(a) above.'; and

Amended under the Supreme Court Act 1981 s. 148(3)(a).

(3) A decision of the High Court under subsection (1) above shall be deemed to be a judgment of the court within the meaning of section 27 of the Supreme Court of Judicature (Consolidation) Act 1925 (appeals to the Court of Appeal), but no appeal shall lie from such a decision unless:

925 c. 49. Amended under the Supreme Court Act 1981 s. 148(3)(b).

(a) the High Court or the Court of Appeal gives leave; and

(b) it is certified by the High Court that the question of law to which its decision relates either is one of general public importance or is one which for some other special reason should be considered by the Court of Appeal.

Exclusion agreements affecting rights under sections 1 and 2.

3. (1) Subject to the following provisions of this section and section 4 below:

(a) the High Court shall not, under section 1(3)(b) above, grant leave to appeal with respect to a question of law arising out of an award, and

(b) the High Court shall not, under section 1(5)(b) above, grant leave to make an application with respect to an award, and

(c) no application may be made under section 2(1)(a) above with respect to a question of law,

if the parties to the reference in question have entered into an agreement in writing (in this section referred to as an 'exclusion agreement') which excludes the right of appeal under section 1 above in relation to that award or, in a case falling within paragraph (c) above, in relation to an award to which the determination of the question of law is material.

(2) An exclusion agreement may be expressed so as to relate to a particular award, to awards under a particular reference or to any other description of awards, whether arising out of the same reference or not; and an agreement may be an exclusion agreement for the purposes of this section whether it is entered into before or after the passing of this Act and whether or not it forms part of an arbitration agreement.

(3) In any case where

(a) an arbitration agreement, other than a domestic arbitration agreement, provides for disputes between the parties to be referred to arbitration, and

(b) a dispute to which the agreement relates involves the question whether a party has been guilty of fraud, and

(c) the parties have entered into an exclusion agreement which is applicable to any award made on the reference of that dispute,

then, except in so far as the exclusion agreement otherwise provides, the High Court shall not exercise its powers under section 24(2) of the principal Act (to take steps necessary to enable the question to be determined by the High Court) in relation to that dispute.

(4) Except as provided by subsection (1) above, sections 1 and 2 above shall have effect notwithstanding anything in any agreement purporting

(a) to prohibit or restrict access to the High Court; or

(b) to restrict the jurisdiction of that court; or

(c) to prohibit or restrict the making of a reasoned award.

(5) An exclusion agreement shall be of no effect in relation to an award made on, or a question of law arising in the course of a reference under, a statutory arbitration, that is to say, such an arbitration as is referred to in subsection (1) of section 31 of the principal Act.

(6) An exclusion agreement shall be of no effect in relation to an award made on, or a question of law arising in the course of a reference under, an arbitration agreement which is a domestric arbitration agreement unless the exclusion agreement is entered into after the commencement of the arbitration in which the award is made or, as the case may be, in which the question of law arises.

(7) In this section 'domestic arbitration agreement' means an arbitration agreement which does not provide, expressly or by implication, for arbitration in a State other than the United Kingdom and to which neither

(a) an individual who is a national of, or habitually resident in, any State other than the United Kingdom, nor

(b) a body corporate which is incorporated in, or whose central management and control is exercised in, any State other than the United Kingdom,

is a party at the time the arbitration agreement is entered into.

Exclusion agreements not to apply in certain cases.

4. (1)Subject to subsection (3) below, if an arbitration award or a question of law arising in the course of a reference relates, in whole or in part, to

(a) a question or claim falling within the Admiralty jurisdiction of the High Court, or

(b) a dispute arising out of a contract of insurance, or

(c) a dispute arising out of a commodity contract,

an exclusion agreement shall have no effect in relation to the award or question unless either

(i) the exclusion agreement is entered into after the commencement of the arbitration in which the award is made or, as the case may be, in which the question of law arises, or

(ii) the award or question relates to a contract which is expressed to be governed by a law other than the law of England and Wales.

(2) In subsection (1)(c) above 'commodity contract' means a contract

(a) for the sale of goods regularly dealt with on a commodity market or exchange in England or Wales which is specified for the purposes of this section by an order made by the Secretary of State; and

(b) of a description so specified.

(3) The Secretary of State may by order provide that subsection (1) above:

(a) shall cease to have effect; or

(b) subject to such conditions as may be specified in the order, shall not apply to any exclusion agreement made in relation to an arbitration award of a description so specified,

and an order under this subsection may contain such supplementary, incidental and transitional provisions as appear to the Secretary of State to be necessary or expedient.

(4) The power to make an order under subsection (2) or

subsection (3) above shall be exercisable by statutory instrument which shall be subject to annulment in pursuance of a resolution of either House of Parliament.

(5) In this section 'exclusion agreement' has the same meaning as in section 3 above.

Interlocutory orders.

5. (1) If any party to a reference under an arbitration agreement fails within the time specified in the order or, if no time is so specified, within a reasonable time to comply with an order made by the arbitrator or umpire in the course of the reference, then, on the application of the arbitrator or umpire or of any party to the reference, the High Court may make an order extending the powers of the arbitrator or umpire as mentioned in subsection (2) below.

(2) If an order is made by the High Court under this section, the arbitrator or umpire shall have power, to the extent and subject to any conditions specified in that order, to continue with the reference in default of appearance or of any other act by one of the parties in like manner as a judge of the High Court might continue with proceedings in that court where a party fails to comply with an order of that court or a requirement of rules of court.

1970 c. 31.

(3) Section 4(5) of the Administration of Justice Act 1970 (jurisdiction of the High Court to be exercisable by the Court of Appeal in relation to judge-arbitrators and judge-umpires) shall not apply in relation to the power of the High Court to make an order under this section, but in the case of a reference to a judge-arbitrator or judge-umpire that power shall be exercisable as in the case of any other reference to arbitration and also by the judge-arbitrator or judge-umpire himself.

(4) Anything done by a judge-arbitrator or judge-umpire in the exercise of the power conferred by subsection (3) above shall be done by him in his capacity as judge of the High Court and have effect as if done by that court.

(5) The preceding provisions of this section have effect notwithstanding anything in any agreement but do not derogate from any powers conferred on an arbitrator or umpire, whether by an arbitration agreement or otherwise.

(6) In this section 'judge-arbitrator' and 'judge-umpire' have the same meaning as in Schedule 3 to the Administration of Justice Act 1970.

6. (1) In subsection (1) of section 8 of the principal Act (agreements where reference is to two arbitrators deemed to include provision that the arbitrators shall appoint an umpire immediately after their own appointment) –

Minor amendments relating to awards and appointment of arbitrators and umpires.

(a) for the words 'shall appoint an umpire immediately' there shall be substituted the words 'may appoint an umpire at any time'; and

(b) at the end there shall be added the words 'and shall do so forthwith if they cannot agree'.

(2) For section 9 of the principal Act (agreements for reference to three arbitrators) there shall be substituted the following section:

'Majority award of three arbitrators.

9. Unless the contrary intention is expressed in the arbitration agreement, in any case where there is a reference to three arbitrators, the award of any two of the arbitrators shall be binding.'

(3) In section 10 of the principal Act (power of court in certain cases to appoint an arbitrator or umpire) in paragraph (c) after the word 'are', in the first place where it occurs, there shall be inserted the words 'required or are' and the words from 'or where' to the end of the paragraph shall be omitted.

(4) At the end of section 10 of the principal Act there shall be added the following subsection:

'(2) In any case where –

(a) an arbitration agreement provides for the appointment of an arbitrator or umpire by a person who is neither one of the parties nor an existing arbitrator (whether the provision applies directly or in default of agreement by the parties or otherwise), and

(b) that person refuses to make the appointment or does not make it within the time specified in the agreement or, if no time is so specified, within a reasonable time,

any party to the agreement may serve the person in question with a written notice to appoint an arbitrator or umpire and, if the appointment is not made within seven clear days after the service of the notice, the High Court or a judge thereof may, on the application of the party who gave the notice, appoint an arbitrator or umpire who shall have the like powers to act in the reference and make an award as if he had been appointed in accordance with the terms of the agreement.'

Application and interpretation of certain provisions of Part I of principal Act.

7. (1) References in the following provisions of Part I of the principal Act to that Part of that Act shall have effect as if the preceding provisions of this Act were included in that Part, namely:

(a) section 14 (interim awards);

(b) section 28 (terms as to costs of orders);

(c) section 30 (Crown to be bound);

(d) section 31 (application to statutory arbitrations); and

(e) section 32 (meaning of 'arbitration agreement').

(2) Subsections (2) and (3) of section 29 of the principal Act shall apply to determine when an arbitration is deemed to be commenced for the purposes of this Act.

(3) For the avoidance of doubt, it is hereby declared that the reference in subsection (1) of section 31 of the principal Act (statutory arbitrations) to arbitration under any other Act does not extend to arbitration under section 92 of the County Courts Act 1959 (cases in which proceedings are to be or may be referred to arbitration) and accordingly nothing in this Act or in Part I of the principal Act applies to arbitration under the said section 92.

1959 c. 22.

Short title, commencement, repeals and extent.

8. (1) This Act may be cited as the Arbitration Act 1979.

(2) This Act shall come into operation on such day as the Secretary of State may appoint by order made by statutory instrument; and such an order

(a) may appoint different days for different provisions of this Act and for the purposes of the operation of the same provision in relation to different descriptions of arbitration agreement; and

(b) may contain such supplementary, incidental and transitional provisions as appear to the Secretary of State to be necessary or expedient.

(3) In consequence of the preceding provisions of this Act, the following provisions are hereby repealed, namely:

(a) in paragraph (c) of section 10 of the principal Act

the words from 'or where' to the end of the paragraph;

(b) section 21 of the principal Act;

(c) in paragraph 9 of Schedule 3 to the Administration of Justice Act 1970, in subparagraph (1) the words '21(1) and (2)' and subparagraph (2). 1970 c. 31.

(4) This Act forms part of the law of England and Wales only.

Appendix E: The Rules of the Supreme Court (Amendment No. 3) 1979

(Showing amendments resulting from the Rules of the Supreme Court (Amendment) 1986.)

STATUTORY INSTRUMENTS

1979 No. 522

SUPREME COURT OF JUDICATURE, ENGLAND AND WALES PROCEDURE

The Rules of the Supreme Court (Amendment No. 3) 1979

Made .... 30 *April* 1979

Laid before Parliament .... 16 *May* 1979

Coming into Operation in accordance with rule 1

We, the Rule Committee of the Supreme Court, being the authority having for the time being power under section 99(4) of the Supreme Court of Judicature (Consolidation) Act 1925* to make, amend or revoke rules regulating the practice and procedure of the Supreme Court of Judicature, hereby exercise those powers and all other powers enabling us in that behalf as follows:

* 1925 c. 49.

Citation and commencement

1. (1) These Rules may be cited as the Rules of the Supreme Court (Amendment No. 3) 1979, and all except Rules 4 to 9, inclusive, shall come into operation on 7 June 1979.

(2) Rules 4 to 9, inclusive, shall come into operation on the day appointed by order for the coming into operation of the Arbitration Act 1979[†] and shall have effect subject to any limitations imposed or provisions made by the order under section 8(2)(a) or (b) of the Act.

(3) In these Rules an Order referred to by number means the Order so numbered in the Rules of the Supreme Court 1965[‡] as amended[§].

Examiners' fees

2. Order 39, Rule 19, shall be amended as follows:

(1) In paragraph (1), in place of the figures £10.00, £2.50 and £17.00 there shall be substituted the figures £20.00, £6.00 and £35.00 respectively.

(2) In paragraph (3), in place of the figures £7.00 and £10.00 there shall be substituted the figures £15.00 and £20.00 respectively.

Deponents' addresses

3. Order 41, Rule 1(4) shall be amended as follows:

(1) After the words 'the first person and' there shall be inserted a comma followed by the words 'unless the Court otherwise directs,'.

(2) At the end of the paragraph there shall be added the following paragraph:

> 'In the case of a deponent who is giving evidence in a professional, business or other occupational capacity the affidavit may, instead of stating the deponent's place of residence, state the address at which he works, the position he holds and the name of his firm or employer, if any.'

Proceedings under the Arbitration Acts

4. Order 59, Rule 14, shall be amended by the addition of the following paragraphs:

> '(5) where an application is made to the Court of Appeal with regard to arbitration proceedings before a judge-arbitrator or judge-umpire which would, in the case of an ordinary arbitrator or umpire, be made to the High Court, the provisions of Order 73, Rule 5, shall apply as if, for the words "the Court", wherever they appear in that rule, there were substituted the words "the Court of Appeal" and as if, for the

[†] 1950 c. 27. [‡] 1975 c. 3.
[§] The relevant amending instruments are SI 1970/944, 1971/1269, 1955, 1977/960, 1955.

words “arbitrator” and “umpire”, there were substituted the words “judge-arbitrator” and “judge-umpire” respectively.

(6) Where an application is made to the Court of Appeal under section 1(5) of the Arbitration Act 1979[*] (including any application for leave), notice thereof must be served on the judge-arbitrator or judge-umpire and on any other party to the reference.’

5. Order 73, Rule 1, shall be amended by deleting the opening words ‘Subject to Order 93, Rule 10(2)(h),’ by inserting after the words ‘Arbitration Act 1950’[†] the words ‘or an appeal or application under the Arbitration Act 1979’ and by inserting after the words ‘the said Act’ the words ‘of 1950’.

6. Order 73, Rule 2, shall be amended as follows:

(1) In paragraph (1), after the words ‘section 23(2) thereof,’ there shall be inserted the following words:

‘or (d) for leave to appeal under section 1(2) of the Arbitration Act 1979, or

(e) to determine, under section 2(1) of that Act, any question of law arising in the course of a reference,’.

(2) In place of the existing paragraph (2) there shall be substituted the following paragraph:

‘(2) Any appeal to the High Court under section 1(2) of the Arbitration Act 1979 shall be made by originating motion to a single judge in court ~~and notice thereof may be included in the notice of application for leave to appeal, where leave is required.’~~

7. Order 73, Rule 3, shall be amended as follows:

(1) In paragraph (1), after the words ‘this Order’, there shall be inserted the words ‘and the provisions of this rule’; and after the words ‘Arbitration Act 1975’[‡] there shall be inserted the words ‘and the Arbitration Act 1979’.

(2) After paragraph (1), in place of the existing paragraphs (2) and (3), there shall be inserted the following paragraphs:

‘(2) Any application under section 1(5) of the Arbitration Act 1979 (including any application for leave), or under section 5 of that Act, shall be made to a judge.

(3) Any application to which this rule applies shall, where an action is pending, be made by summons in the action, and in any other case by an originating summons for which no appearance need be entered.

(4) Where an application is made under section 1(5) of the Arbitration Act 1979 (including any application for leave), the

---

[*] 1979 c. 42. [†] 1950 c. 27. [‡] 1975 c. 3.

summons must be served on the arbitrator or umpire and on any other party to the reference.'

8. In Order 73, in place of the existing Rules 5 and 6 there shall be inserted the following rule:

'Time-limits and other special provisions as to appeals and applications under the Arbitration Acts

5. (1) An application to the Court

(a) to remit an award under section 22 of the Arbitration Act 1950, or

(b) to set aside an award under section 23(2) of that Act or otherwise, or

(c) to direct an arbitrator or umpire to state the reasons for an award under section 1(5) of the Arbitration Act 1979,

must be made, and the summons or notice must be served, within 21 days after the award has been made and published to the parties.

(2) In the case of an appeal to the Court under section 1(2) of the Arbitration Act 1979, the summons for leave to appeal, where leave is required, and the notice of originating summons must be served, and the appeal entered, within 21 days after the award has been made and published to the parties:

Provided that, where reasons material to the appeal are given on a date subsequent to the publication of the award, the period of 21 days shall run from the date on which the reasons are given.

(3) An application, under section 2(1) of the Arbitration Act 1979, to determine any question of law arising in the course of a reference, must be made, and notice thereof served, within 14 days after the arbitrator or umpire has consented to the application being made, or the other parties have so consented.

(4) For the purpose of paragraph (3) the consent must be given in writing.

(5) In the case of every appeal or application to which this rule applies, the notice of originating motion, the originating summons or the sumons, as the case may be must state the grounds of the appeal or application and, where the appeal or application is founded on evidence by affidavit, or is made with the consent of the arbitrator or umpire or of the other parties, a copy of every affidavit intended to be used, or, as the case may be, of every consent given in writing, must be served with that notice.

Applications and appeals to be heard by Commercial Judges

6. (1) Any matter which is required, by Rule 2 or 3, to be heard by a judge, shall be heard by a Commercial Judge, unless any such judge otherwise directs.

(2) Nothing in the foregoing paragraph shall be construed as preventing the powers of a Commercial Judge from being exercised by any judge of the High Court.'

9. Subparagraph (h) of Order 93, Rule 10(2), shall be omitted.

Proceedings concerning the International Oil Pollution Compensation Fund

10. Order 75 shall be amended as follows:

(1) In Rule 2(1), the word 'and' at the end of subparagraph (a) shall be deleted and the following words shall be inserted after subparagraph (b):

'and (c) every action to enforce a claim under section 1 of the Merchant Shipping (Oil Pollution) Act 1971* or section 4 of the Merchant Shipping Act 1974†.'

(2) After Rule 2, there shall be inserted the following rule:

'*Proceedings against, or concerning, the International Oil Pollution Compensation Fund*

2A. (1) All proceedings against the International Oil Pollution Compensation Fund (in this rule referred to as "the Fund") under section 4 of the Merchant Shipping Act 1974 shall be commenced in the Admiralty Registry.

(2) For the purposes of section 6(2) of the Merchant Shipping Act 1974, any party to proceedings brought against an owner or guarantor in respect of liability under section 1 of the Merchant Shipping (Oil Pollution) Act 1971 may give notice to the Fund of such proceedings by serving a notice in writing on the Fund together with a copy of the writ and copies of the pleadings (if any) served in the action.

(3) The Court shall, on the application made *ex parte* by the Fund, grant leave to the Fund to intervene in any proceedings to which the preceding paragraph applies, whether notice of such proceedings has been served on the Fund or not, and paragraphs (3) and (4) of Rule 17 shall apply to such an application.

(4) Where judgment is given against the Fund in any proceedings under section 4 of the Merchant Shipping Act 1974, the registrar shall cause a stamped copy of the judgment to be sent by post to the Fund.

(5) The Fund shall notify the registrar of the matters set out in section 4(12)(b) of the Merchant Shipping Act 1974 by a notice in writing, sent by post to, or delivered at, the registry.'

---

* 1971 c. 59.

† 1974 c. 43.

Summary proceedings for possession of land

11. Order 113, Rule 2, shall be amended by substituting the following words for the existing three paragraphs:

> 'The originating summons shall be in Form No. 11A in Appendix A and no appearance need be entered to it.'

12. In Order 113, Rule 3(c), the words 'where the summons is in Form No. 11A,' shall be omitted.

13. Order 113, Rule 4(2), shall be amended as follows:

(1) For the words 'A summons in Form No. 11A' there shall be substituted the words 'The summons'.

(2) After the words 'a copy of the summons' in subparagraphs (a) and (b) there shall be inserted the words 'and a copy of the affidavit'.

Elwyn-Jones, C.
Widgery, C.J.
Denning, M.R.
George Baker, P.
Eustace Roskill, L.J.
R. E. Megarry, V-C.
Hilary Talbot, J.
Patrick O'Connor, J.
J. Maurice Price.
John Toulmin.
H. Montgomery-Campbell.
Harold Hewitt.

Dated 30 April 1979

EXPLANTORY NOTE

(*This Note is not part of the Rules.*)

These Rules amend the Rules of the Supreme Court so as –

(a) to raise the fees payable to examiners of the Court (Rule 2);
(b) to allow a person making an affidavit to give a 'work address' instead of a 'home address' in certain circumstances and also to apply for permission to omit the address altogether (Rule 3);
(c) to make provision for proceedings under the Arbitration Act 1979 once the 'case stated' procedure under section 21 of the Arbitration Act 1950 has ceased to be available (Rules 4 to 9);
(d) to make provision for proceedings affecting the International Oil Pollution Compensation Fund (Rule 10); and
(e) to amend the rules applicable to summary proceedings for possession of land so as to require Form No. 11A to be used in every case (Rules 11 to 13).

Appendix F: The Arbitration Act 1979 (Commencement) Order 1979

STATUTORY INSTRUMENTS

1979 No. 750 (C.16)

ARBITRATION

The Arbitration Act 1979 (Commencement) Order 1979

Made .... 28 *June* 1979

The Secretary of State in exercise of the powers conferred on him by section 8(2) of the Arbitration Act 1979* hereby makes the following Order:

Citation and interpretation

1. (1) This Order may be cited as the Arbitration Act 1979 (Commencement) Order 1979.

(2) In this Order 'the Act' means the Arbitration Act 1979.

Appointed day

2. The Act shall come into operation on 1 August 1979 (hereinafter referred to as 'the appointed day'), but, except as provided in Article 3 of this Order, shall not apply to arbitrations commenced before that date.

3. If all the parties to a reference to arbitration commenced before the appointed day have agreed in writing that the Act should apply to that arbitration, the Act shall so apply from the appointed day or the date of the agreement whichever is the later.

Cecil Parkinson
Minister of State
Department of Trade

28 June 1979

* 1979 c. 42.

Explanatory note

(This Note is not part of the Order.)

This order appoints 1 August 1979 as the day on which the Arbitration Act 1979 ('the Act') comes into operation ('the appointed day').

The Act provides a new system of judicial review of arbitration awards and repeals section 21 of the Arbitration Act 1950 (c. 27) which relates to a statement of a special case for a decision of the High Court, but the Order provides that the Act does not apply to an arbitration commenced before the appointed day, unless the parties agree that it should. Under section 29(2) of the 1950 Act an arbitration is deemed to be commenced when a notice is served by a party to an arbitration agreement requiring the appointment of an arbitrator or requiring the other parties to submit a dispute to an arbitrator designated by the agreement.

Appendix G: Interest tables

Clause 60(6) of the ICE Conditions, fifth edition, and Clause 60(7) of the sixth edition, provide for payment to the contractor of interest on overdue payments, at the rate of 2% over base rate, compounded monthly (see Chapter 8). Similar provision is made in Clause 15(3)(f) of the FCEC Form of Subcontract, wherein the subcontractor is expressly entitled to interest on overdue sums *at the rate payable . . . under the provisions of the main contract.*There is no corresponding provision under the JCT form of building contract.

Section 19A of the 1950 Act empowers an arbitrator, at his discretion, to award *simple* interest at a rate which is also at his discretion, on any sum which he awards or on any sum which is the subject of the reference but which is paid before the award, for such period ending not later than the date of the award as he thinks fit. The arbitrator must however exercise his discretion judicially: in many cases the rate of interest used in calculating the amount due is taken to be 2% over base rate; and the starting date of the interest period is taken to be *the date on which the money ought to have been paid.*

It follows that the main difference between interest awarded under the relevant term of the (ICE) contract and that awarded under the statutory power, is that interest under the former provision is compounded monthly, while under the latter it must be simple.

Table 1 provides a factual record of changes in base rate from 1 January 1986 until the publication date of this book, together with mean base rates plus 2% for each month.

Table 2 gives factors for calculating either simple or compound interest at those rates. A month is taken as being one-twelfth of a year irrespective of the number of days in the month.

USE OF TABLE 2 FOR SIMPLE AND COMPOUND INTEREST CALCULATIONS

1. Factors FS (simple) and FC (compound) in Table 2 relate to the *end* of each month. Factors for the first day of each month are those tabulated for the end of the previous month.
2. For simple interest calculations, interpolate to find factors FS1 and FS2 on dates of start and end of interest period, using tabulated factors FS for end of previous month and adding the relevant proportion of the month's factor.
3. Then simple interest = Principal sum × (FS2 – FS1).
4. For compound interest calculations, factors FC1 and FC2 are those tabulated for the end of the previous month, multiplied by (1 plus the relevant proportion of the month factor minus 1).
5. Then compound interest = Principal sum × (FC2/FC1 – 1).

Example

Interest on £3.6m for period from 11.06.86 to 31.12.92:

Simple interest:

FS1 = 0.0567728 + 11/30 × 0.0100000 = 0.0604395
FS2 = 0.9459054

Interest = £3 600 000 × (0.9459054 – 0.0604395) = £3 187 677

Compound interest:
FC1 = 1.0580756 × (1 + 11/30 × 0.0100000) = 1.0619552
FC2 = 2.5612169

Interest = £3 600 000 × (2.5612169/1.0619552 – 1) = £5 082 457

Table 1 Record of base rates and monthly mean rates + 2%

From	*Days*	*Base %*	*M mean + 2%*
01.01.86.	8	11.5	
09.01.86.	23	12.5	14.2419355
01.02.86.	28	12.5	14.5000000
01.03.86.	18	12.5	
19.03.86.	13	11.5	14.0806452
01.04.86.	7	11.5	
08.04.86.	13	11.0	
21.04.86.	10	10.5	12.9500000
01.05.86.	22	10.5	
23.05.86.	9	10.0	12.3548387
01.06.86.	30	10.0	12.0000000
01.07.86.	31	10.0	12.0000000
01.08.86.	31	10.0	12.0000000
01.09.86	30	10.0	12.0000000
01.10.86.	13	10.0	
14.10.86.	18	11.0	12.5806452
01.11.86.	30	11.0	13.0000000
01.12.86.	31	11.0	13.0000000
01.01.87.	31	11.0	13.0000000
01.02.87.	28	11.0	13.0000000
01.03.87.	9	11.0	
10.03.87.	8	10.5	
18.03.87.	14	10.0	12.4193548
01.04.87.	27	10.0	
28.04.87.	3	9.5	11.9500000
01.05.87.	10	9.5	
11.05.87.	21	9.0	11.1612903
01.06.87.	30	9.0	11.0000000
01.07.87.	31	9.0	11.0000000
01.08.87.	6	9.0	
07.08.87.	25	10.0	11.8064516
01.09.87.	30	10.0	12.0000000
01.10.87.	24	10.0	
25.10.87.	7	9.5	11.8870968
01.11.87.	4	9.5	
05.11.87.	26	9.0	11.0666667
01.12.87.	3	9.0	
04.12.87.	28	8.5	10.5483871
01.01.88.	31	8.5	10.5000000
01.02.88.	1	8.5	
02.02.88.	28	9.0	10.9827586
01.03.88.	17	9.0	
18.03.88.	14	8.5	10.7741935

Table 1 *Continued*

From	*Days*	*Base %*	*M mean + 2%*
01.04.88.	10	8.5	
11.04.88.	20	8.0	10.1666667
01.05.88.	17	8.0	
18.05.88.	14	7.5	9.7741935
01.06.88.	6	7.5	
07.06.88.	16	8.5	
23.06.88.	6	9.0	
29.06.88.	2	9.5	10.4666667
01.07.88.	4	9.5	
05.07.88.	14	10.0	
19.07.88.	13	10.5	12.1451613
01.08.88.	8	10.5	
09.08.88.	17	11.0	
26.08.88.	6	12.0	13.0645161
01.09.88.	30	12.0	14.0000000
01.10.88.	31	12.0	14.0000000
01.11.88.	24	12.0	
25.11.88.	6	13.0	14.2000000
01.12.88.	31	13.0	15.0000000
01.01.89.	31	13.0	15.0000000
01.02.89.	28	13.0	15.0000000
01.03.89.	31	13.0	15.0000000
01.04.89.	30	13.0	15.0000000
01.05.89.	20	13.0	
21.05.89.	11	14.0	15.3548387
01.06.89.	30	14.0	16.0000000
01.07.89.	31	14.0	16.0000000
01.08.89.	31	14.0	16.0000000
01.09.89.	30	14.0	16.0000000
01.10.89.	4	14.0	
05.10.89.	27	15.0	16.8709677
01.11.89.	30	15.0	17.0000000
01.12.89.	31	15.0	17.0000000
01.01.90.	31	15.0	17.0000000
01.02.90.	28	15.0	17.0000000
01.03.90.	31	15.0	17.0000000
01.04.90.	30	15.0	17.0000000
01.05.90.	31	15.0	17.0000000
01.06.90.	30	15.0	17.0000000
01.07.90.	31	15.0	17.0000000
01.08.90.	31	15.0	17.0000000
01.09.90.	30	15.0	17.0000000

Table 1 *Continued*

From	*Days*	*Base %*	*M mean + 2%*
01.10.90.	5	15.0	
06.10.90.	26	14.0	16.1612903
01.11.90.	30	14.0	16.0000000
01.12.90.	31	14.0	16.0000000
01.01.91.	31	14.0	16.0000000
01.02.91.	11	14.0	
12.02.91.	15	13.5	
27.02.91.	2	13.0	15.6607143
01.03.91.	21	13.0	
22.03.91.	10	12.5	14.8387097
01.04.91.	11	12.5	
12.04.91.	19	12.0	14.1833333
01.05.91.	23	12.0	
24.05.91.	8	11.5	13.8709677
01.06.91.	30	11.5	13.5000000
01.07.91.	12	11.5	
13.07.91.	19	11.0	13.1935484
01.08.91.	31	11.0	13.0000000
01.09.91.	3	11.0	
04.09.91.	27	10.5	12.5500000
01.10.91.	31	10.5	12.5000000
01.11.91.	30	10.5	12.5000000
01.12.91.	31	10.5	12.5000000
01.01.92.	31	10.5	12.5000000
01.02.92.	29	10.5	12.5000000
01.03.92	31	10.5	12.5000000
01.04.92	30	10.5	12.5000000
01.05.92	4	10.5	
05.05.92	27	10.0	12.0645161
01.06.92	30	10.0	12.0000000
01.07.92	31	10.0	12.0000000
01.08.92	31	10.0	12.0000000
01.09.92	16	10.0	
17.09.92	1	12.0	
18.09.92	5	10.0	
23.09.92	8	9.0	11.8000000
01.10.92	16	9.0	
17.10.92	15	8.0	10.5161290
01.11.92	12	8.0	
13.11.92	18	7.0	9.4000000
01.12.92	31	7.0	9.0000000
01.01.93	26	7.0	
27.01.93	5	6.0	8.8387097
01.02.93	28	6.0	8.0000000
01.03.93			

Table 2 Monthly interest at base rate + 2%: (FS) simple; (FC) compounded

Month	*M mean + 2%*	*Mnth factor*	*Factor FS*	*Factor FC*
JAN 86	14.2419355	1.0118683	0.0118683	1.0118683
FEB 86	14.5000000	1.0120833	0.0239516	1.0240950
MAR 86	14.0806452	1.0117339	0.0356855	1.0361116
APR 86	12.9500000	1.0107917	0.0464772	1.0472930
MAY 86	12.3548387	1.0102957	0.0567728	1.0580756
JUN 86	12.0000000	1.0100000	0.0667728	1.0686564
JUL 86	12.0000000	1.0100000	0.0767728	1.0793429
AUG 86	12.0000000	1.0100000	0.0867728	1.0901364
SEP 86	12.0000000	1.0100000	0.0967728	1.1010377
OCT 86	12.5806452	1.0104839	0.1072567	1.1125809
NOV 86	13.0000000	1.0108333	0.1180901	1.1246338
DEC 86	13.0000000	1.0108333	0.1289234	1.1368173
JAN 87	13.0000000	1.0108333	0.1397567	1.1491329
FEB 87	13.0000000	1.0108333	0.1505901	1.1615818
MAR 87	12.4193548	1.0103495	0.1609395	1.1736036
APR 87	11.9500000	1.0099583	0.1708978	1.1852907
MAY 87	11.1612903	1.0093011	0.1801989	1.1963152
JUN 87	11.0000000	1.0091667	0.1893656	1.2072814
JUL 87	11.0000000	1.0091667	0.1985323	1.2183481
AUG 87	11.8064516	1.0098387	0.2083710	1.2303351
SEP 87	12.0000000	1.0100000	0.2183710	1.2426385
OCT 87	11.8870968	1.0099059	0.2282769	1.2549479
NOV 87	11.0666667	1.0092222	0.2374991	1.2665213
DEC 87	10.5483871	1.0087903	0.2462894	1.2776545
JAN 88	10.5000000	1.0087500	0.2550394	1.2888339
FEB 88	10.9827586	1.0091523	0.2641917	1.3006297
MAR 88	10.7741935	1.0089785	0.2731702	1.3123074
APR 88	10.1666667	1.0084722	0.2816424	1.3234256
MAY 88	9.7741935	1.0081452	0.2897876	1.3342051
JUN 88	10.4666667	1.0087222	0.2985098	1.3458423
JUL 88	12.1451613	1.0101210	0.3086308	1.3594636
AUG 88	13.0645161	1.0108871	0.3195179	1.3742642
SEP 88	14.0000000	1.0116667	0.3311846	1.3902973
OCT 88	14.0000000	1.0116667	0.3428512	1.4065174
NOV 88	14.2000000	1.0118333	0.3546846	1.4231612
DEC 88	15.0000000	1.0125000	0.3671846	1.4409507
JAN 89	15.0000000	1.0125000	0.3796846	1.4589626
FEB 89	15.0000000	1.0125000	0.3921846	1.4771996
MAR 89	15.0000000	1.0125000	0.4046846	1.4956646
APR 89	15.0000000	1.0125000	0.4171846	1.5143604
MAY 89	15.3548387	1.0127957	0.4299803	1.5337377
JUN 89	16.0000000	1.0133333	0.4433136	1.5541876
JUL 89	16.0000000	1.0133333	0.4566469	1.5749101
AUG 89	16.0000000	1.0133333	0.4699803	1.5959089
SEP 89	16.0000000	1.0133333	0.4833136	1.6171876
OCT 89	16.8709677	1.0140591	0.4973727	1.6399239
NOV 89	17.0000000	1.0141667	0.5115394	1.6631562
DEC 89	17.0000000	1.0141667	0.5257061	1.6867175
JAN 90	17.0000000	1.0141667	0.5398727	1.7106127
FEB 90	17.0000000	1.0141667	0.5540394	1.7348464
MAR 90	17.0000000	1.0141667	0.5682061	1.7594234
APR 90	17.0000000	1.0141667	0.5823727	1.7843486
MAY 90	17.0000000	1.0141667	0.5965394	1.8096268
JUN 90	17.0000000	1.0141667	0.6107061	1.8352632
JUL 90	17.0000000	1.0141667	0.6248727	1.8612628
AUG 90	17.0000000	1.0141667	0.6390394	1.8876307
SEP 90	17.0000000	1.0141667	0.6532061	1.9143721
OCT 90	16.1612903	1.0134677	0.6666738	1.9401544

Table 2 *Continued*

Month	*M mean + 2%*	*Mnth factor*	*Factor FS*	*Factor FC*
NOV 90	16.0000000	1.0133333	0.6800071	1.9660231
DEC 90	16.0000000	1.0133333	0.6933405	1.9922367
JAN 91	16.0000000	1.0133333	0.7066738	2.0187999
FEB 91	15.6607143	1.0130506	0.7197244	2.0451464
MAR 91	14.8387097	1.0123656	0.7320900	2.0704359
APR 91	14.1833333	1.0118194	0.7439094	2.0949073
MAY 91	13.8709677	1.0115591	0.7554686	2.1191226
JUN 91	13.5000000	1.0112500	0.7667186	2.1429627
JUL 91	13.1935484	1.0109946	0.7777132	2.1665238
AUG 91	13.0000000	1.0108333	0.7885465	2.1899945
SEP 91	12.5500000	1.0104583	0.7990049	2.2128982
OCT 91	12.5000000	1.0104167	0.8094215	2.2359492
NOV 91	12.5000000	1.0104167	0.8198382	2.2592403
DEC 91	12.5000000	1.0104167	0.8302549	2.2827741
JAN 92	12.5000000	1.0104167	0.8406715	2.3065530
FEB 92	12.5000000	1.0104167	0.8510882	2.3305796
MAR 92	12.5000000	1.0104167	0.8615049	2.3548564
APR 92	12.5000000	1.0104167	0.8719215	2.3793862
MAY 92	12.0645161	1.0100538	0.8819753	2.4033080
JUN 92	12.0000000	1.0100000	0.8919753	2.4273410
JUL 92	12.0000000	1.0100000	0.9019753	2.4516145
AUG 92	12.0000000	1.0100000	0.9119753	2.4761306
SEP 92	11.8000000	1.0098333	0.9218086	2.5004792
OCT 92	10.5161290	1.0087634	0.9305721	2.5223920
NOV 92	9.4000000	1.0078333	0.9384054	2.5421508
DEC 92	9.0000000	1.0075000	0.9459054	2.5612169
JAN 93	8.8387097	1.0073656	0.9532710	2.5800818
FEB 93	8.0000000	1.0066667	0.9599377	2.5972823

Appendix H: Flow chart

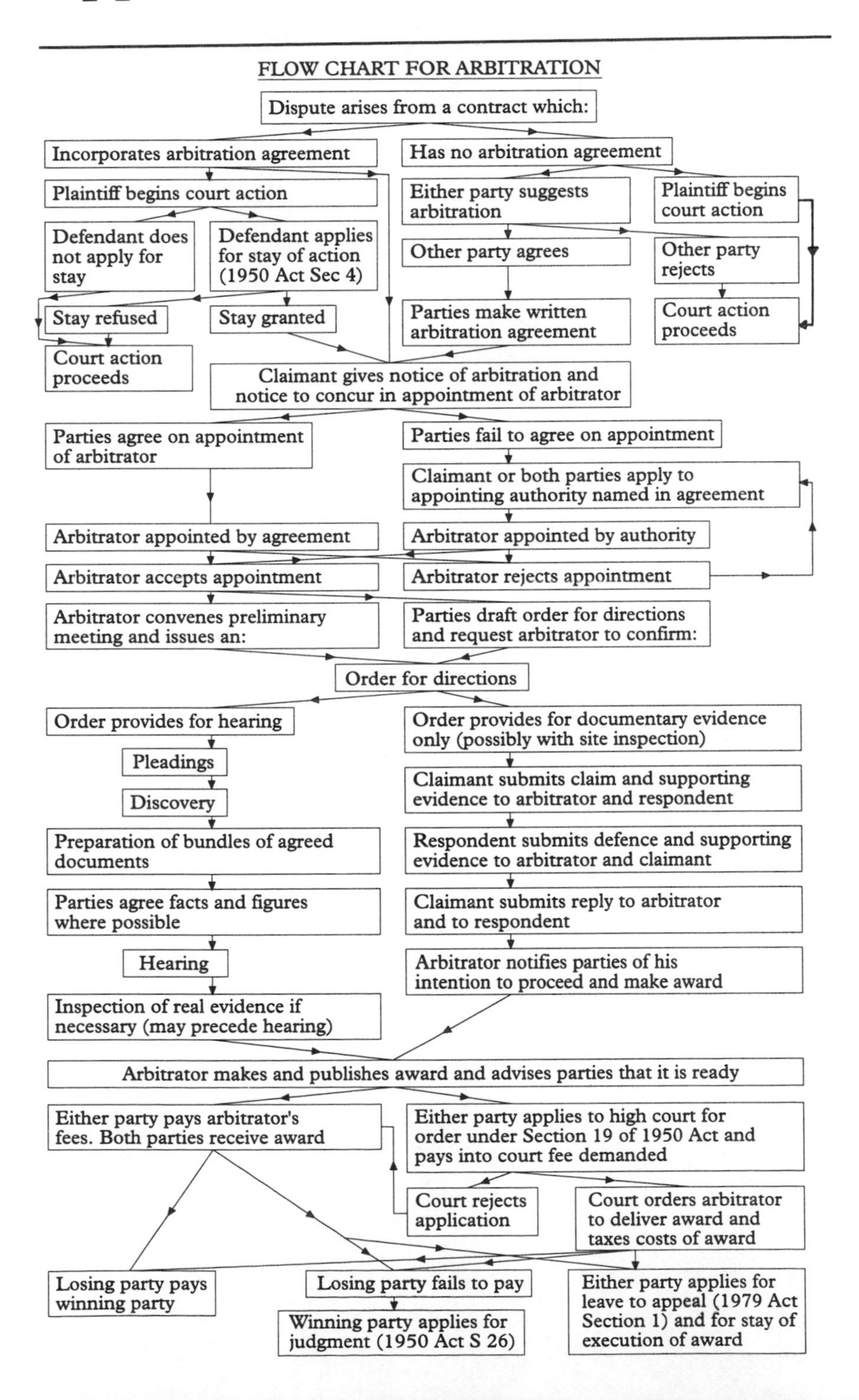

Bibliography

ARBITRATION

Mustill and Boyd, *Commercial Arbitration*, 2nd edn, Butterworths.
Walton and Vitoria, *Russell on Arbitration*, 20th edn, Stevens.
Bernstein, R., *Handbook of Arbitration Practice*, Sweet & Maxwell.
Hawker, Uff and Timms, *The Institution of Civil Engineers' Arbitration Practice*, Thomas Telford.
Marshall, *Gill: The Law of Arbitration*, 3rd edn, Sweet & Maxwell.

CONSTRUCTION CONTRACT LAW

Abrahamson, Max W., *Engineering Law and the ICE Contracts*, 4th edn, Applied Science Publishers.
Duncan Wallace, I., *Hudson's Building & Engineering Contracts*, 10th edn, Sweet & Maxwell.
Powell-Smith and Stephenson, *Civil Engineering Claims*, 2nd edn, BSP Professional Books.
Powell-Smith and Furmston, *A Building Contract Casebook*, Granada.
Powell-Smith and Chappell, *Building Contract Dictionary*, Architectual Press.

CONTRACT LAW IN GENERAL

Furmston, M. P., *Cheshire and Fifoot's Law of Contract*, Butterworths.
Schmitthoff and Sarre, *Charlesworth's Mercantile Law*, 14th edn, Stevens.

LAW IN GENERAL

Eddy, K. J., *The English Legal System*, Sweet & Maxwell.
Bird, R., *Osborne's Concise Law Dictionary*, 7th edn, Sweet & Maxwell.

Table of Cases

Note: BLR corresponds to Building Law Report, TLR to Times Law Report and WLR to Weekly Law Report.

Index